Cognitive Structure and Development in Nonhuman Primates

COMPARATIVE COGNITION
AND NEUROSCIENCE

Thomas G. Bever, David S. Olton,
and Herbert L. Roitblat, Senior Editors

Cognitive Structure and Development in Nonhuman Primates

Edited by

Francesco Antinucci

Istituto di Psicologia, C.N.R.
Rome, Italy

Psychology Press
Taylor & Francis Group

New York London

First Published by Lawrence Erlbaum Associates, Inc., Publishers
10 Industrial Avenue
Mahwah, New Jersey 07430

Reprinted 2010 by Psychology Press

Lawrence Erlbaum Associates, Inc., Publishers
365 Broadway
Hillsdale, New Jersey 07642

Library of Congress Cataloging in Publication Data

Cognitive structure and development in nonhuman primates / edited by
    Francesco Antinucci.
        p.   cm. — (Comparative cognition and neuroscience)
    Bibliography: p.
    Includes indexes.
    ISBN 0-8058-0242-8. — ISBN 0-8058-0544-3 (pbk.)
    1. Primates—Behavior. 2. Cognition in animals. I. Antinucci,
Francesco.
    [DNLM: 1. Cognition. 2. Primates—growth & development.   QL
737.P9   C676]
QL737.P9C54   1989
599.8′0451—dc20
DNLM/DLC
for Library of Congress   89-11670
                                                                     CIP

10   9   8   7   6   5   4   3   2

Table of Contents

Preface

About ten years ago, quite casually, I happened to follow an infant monkey during its early development. Up to that moment, I had been studying language capacities and, especially, their development in children and their relation to cognitive development. I was struck by the intriguing pattern of similarities and differences that this animal's developing cognition displayed in comparison to that of the human infant: Sometimes I could predict exactly which behavior was about to appear; other times I was simply baffled by its unusual developments. But, to me, the most important experience was seeing the dissociation of abilities whose coherent development and "natural" basis I had always taken for granted.

I was puzzled by this problem and, at the same time, attracted by its potentiality to reveal the deeper nature of the organization of cognitive capacities.

At first I thought I could just consult the animal literature and find out, if not the answers, at least the relevant data (like a description of the cognitive development of various species of primates). Not only did I not find anything like that, but I could hardly make sense of what this literature was about. My original background had been in Piagetian psychology, followed (I should say, without feeling much discontinuity) by the cognitive psychology of the late sixties. I couldn't find anything even distantly related to the study of cognitive capacities as they were being, at that time, investigated in man.

Thus, if I wanted the right data, I had to collect my own. The idea was simple enough: follow the development of infant nonhuman primates and test their cognitive achievements in the same way as I would do with human infants.

From that moment on, a collective enterprise started; in fact, a much larger and longer one than I could have anticipated. Contrary to the research practice I was accustomed to, I couldn't just go out and test my subjects; in this case, I had to house, raise, support, and take care of them. Several people were involved, at all levels and at all phases, over the years: Those who endured following the original program through its many difficulties, and did not lose their patience at pacing the satisfaction of their intellectual curiosity to the rhythm of each of our developing subjects, have authored the chapters of this book.

But this book would not be here without the help, at one time or another, of the many more who, from night nursing to babysitting, from patiently testing to tediously transcribing, from analyzing data to arguing our theses, made it possible. To all of them go our thanks: Luisa Berlinguer, Anna Carnera, Patrizia Costa, Andrea Misiti, Ageliki Nicolopoulou, Domitilla Nonis, Sue Parker, Maria Cristina Riviello, Gabriele Schino, Guido Travaglia, Loredana Trinca, Anna Troise, and Elisabetta Visalberghi. No less important was the help of two friends who, in different ways, supported our work all along. Raffaello Misiti, the former director of our institute, instead of balking at this unorthodox research proposal, lent to it his trust and support wholeheartedly. Unfortunately, he is no longer with us today to see its long-awaited fruits. To Jonas Langer, we owe in more than one way. He showed us how to make sense of the apparently chaotic and haphazard action of infants: Without his work on the cognition of the human infant, there would have been for us no basis of comparison. He freely gave us his time, intellectual wit, and encouragement, in Berkeley and in Rome. But most of all, he offered us the example of his personal coherence and perseverance in pursuing a lasting research goal. It is a pleasure to have his chapter in this book.

Francesco Antinucci

I INTRODUCTION

1 The Comparative Approach to the Study of Cognition

Francesco Antinucci
Istituto di Psicologia, C.N.R.
Rome, Italy

The studies presented in this book grew out of a line of research attempting to pursue the most classical aims and themes of comparative psychology as it was originally established and formulated, at the turn of the century, in the works of Romanes, Hobhouse, Kohler, etc. Unfortunately, with very few exceptions, they are not (and have not been for a long time) the aims and themes of what is still today recognized, taught and practiced as comparative psychology. Hence, the need of this introduction that would have been superfluous had it still been true that "[Comparative Psychology] . . . is concerned with the genesis of the human mind as such. It seeks to determine the stages of development which lead from the first beginning of psychic life to the emergence of the human reason" (Hobhouse, 1968, p.117). As recently remarked by Tolman (1987, p. 288), "The vision of comparative psychology expressed by Romanes, Morgan, and Hobhouse was one of a distinctly human psychology, charged with identifying the unique qualities of human psyche (or mental functioning), and explaining them in terms of their evolution". If its object is the evolutionary study of the human mind, then, in purely darwinian terms, "comparative" refers to the classical method that allows, as in morphology or anatomy, the determination and reconstruction of the evolutionary path through the systematic comparison of corresponding structures in increasingly removed, related species. In the specific case, the "minds" of other species, beginning with man's closest relatives. Kohler writes, in the introductory justification of his study of the intellectual capacities of apes:

3

> Two sets of interests lead us to test the intelligence of higher apes. We are
> aware that it is a question of beings which in many cases are nearer to man
> than to the other ape species;. . . .We wished to ascertain the degree of rela-
> tionship between anthropoid apes and man in a field which seems to us partic-
> ularly important, but on which we have as yet little information. (1976, p. 1)

As we shall see in a short while, no less important is the second "set of
interests" motivating his choice.

Of the three key concepts defining, as we have just seen, the comparative
psychology enterprise (*evolutionary* study of the *human mind*) not one
survived the behavioristic revolution: not "evolutionary through the comparative method", not "human" and, especially, not "mind". Through an
historical process that has recently been amply documented (see, among
many others, Dewsbury, 1984; Epstein, 1987, O'Donnell, 1985), comparative psychology became the "synchronic" study of the behavior of one or
two species of animals. By 1927, another one of the "founding fathers",
attending him too, as Kohler and for the same reasons, to the study of the
intellectual capacities of man's closest relatives, R. Yerkes, felt obliged to
write in the introduction to one of his classical monographs:

> As a fact, the term mind (and in the present title mental) is used to protest
> against what he [the author] deems unwarranted and unprofitable assump-
> tions of the cult of behaviorism and as indication that he considers mental,
> no less than behavioral phenomena, material of biological science. (Yerkes,
> 1927b, p. 1)

The title of Yerkes' monograph was "The mind of a gorilla" and the subtitle
of its second part, in the introduction to which the statement quoted above
appears, "Mental development".

This is not the place to rerun the complex historical and epistemological
process that led from the just preoccupations with inferences of mind
functioning based on analogical extrapolation of introspective results
("Starting from what I know of the operations of my own individual mind,
and the activities which in my own organism they prompt, I proceed by
analogy to infer from the observable activities of other organisms what are
the mental operation that underlie them", Romanes, 1883, pp.1–2) to the
curious means-end distortion thereby overt behavior became the direct end
of the investigation, rather than the indirect mean to access mental functioning (see Wasserman, 1984).

One point should, however, be mentioned because it is directly relevant
to the resurrection of the original enterprise, and to the approach presented
here. Quite independently from its physicalist methodological and theoretical restrictions (which, in fact, were to be progressively relaxed in the

60 years of its history), behaviorism espoused a peculiar metatheoretical assumption (and a recurrent one in the history of several disciplines, see Antinucci, 1982, for a wider discussion), viz., that the whole of cognitive structure and functioning could be exhaustively accounted for in terms of one simple atomic mechanism, the association of good empiricist memory. Obvious macroscopic differences among levels of structure and functioning, as well as among species capacities are, in principle, conceived of as being of quantitative nature: more easily established, more extensive, richer and longer, associations and associations of associations account for increasing complexity (see what Roitblat, 1987, pp.54–55, appropriately terms the "ubiquity" and "equipotentiality" assumptions of behaviorism). Given this assumption, what is needed most of all is a thorough understanding of all the parameters regulating the basic mechanism (reinforcement, extinction, generalization, differentiation, etc.): complexity would then result from some simple summation function. Hence, the rat or the pigeon became ideal object of study, not, however, as terms of an extended evolutionary comparison with man, according to the dictate of the original program, but as valuable, because essentially isomorphic, small-scale models of larger organizations. In fact, the complexity-as-quantity assumption thoroughly trivializes the comparative function and, at the same time and for the same reasons, rescues from the realm of biological absurdities the possibility of directly relating even two species belonging to two different taxa, like pigeon and man: "Pigeon, rat, monkey, which is which? It doesn't matter" (Skinner, 1956, p.230).

Thus, "mind" disappeared for epistemological reasons, "human" became interchangable with animal for theoretical ones and "evolutionary" was emptied of meaning. No wonder that when the study of human cognition was put back on its feet, in the sixties, the comparative approach, turned into the study of animal learning behavior, had very little to say. The two fields drifted more and more apart, and while the first was developing increasingly more sophisticated models of the mind "software", the second gravitated more and more toward the physiological "hardware", as testified, for example, by the distribution of articles appearing in the *Journal of Comparative and Physiological Psychology*.

The repeal of the atomistic assumption is at the core of the "cognitive revolution" (see Chomsky, 1959; Miller, Galanter, & Pribram, 1960). With it there came about a return (obviously, on richer and more sophisticated bases) to structural conceptions. Cognitive capacities are best viewed as integrated structures, that is to say, they cannot be reduced to the simple summation of independent atomic constituents. This is not to say that they are some kind of metaphysical indecomposable whole, of organicistic or vitalistic memory, but simply that their specific properties and functioning depend crucially on the reciprocal relations and delicately orchestrated interactions of their constituent elements, rather than on their elementary nature. To take an obvious, but clear example, human language is not a list of words, and sentences that instantiate its functioning are not sums of juxtaposed words.

It is this network of interdependences and its crucial role in functioning that is referred to by the word "structure". Because of their very organization, structures tend to show a number of distinct properties: scale differences translate into qualitative differences, i. e., show novel properties, levels of functioning are irreducible to each other, and, especially, causal determinants tend to have non-linear effects ("trigger" causation).

Owing to these properties, model-building within this framework is a much more complicated task. Interdependence of parts, non-scaling and trigger causation cooperate in rendering extremely indirect the relation between external events observed in functioning and underlying processes generating them. This creates two hard, related problems to the theorist: arbitrariness of domain delimitation, and model underdetermination. How are we to delimit underlying unitary mechanisms in largely integrated systems on a non-arbitrary basis? External phenomena are, as we have just seen, no criterial guide because no one-to-one correspondence can be assumed between them and underlying mechanism. What appears, for example, as language behavior can be determined by a host of separate interacting mechanisms, rather than by a unitary, self-suffcent language capacity; and conversely, each of these hypothetical mechanisms (like, for example, a single processor for serial units) might play crucial roles in generating a variety of apparently unrelated behaviors (see, for example, Bever, 1974).

Or, to take an example closer to the topics that will be dealt with in this book, should we pool tool-use behaviors together as corresponding to a proper domain of cognition (as done by many students), or the capacity that makes it possible to use a stick to rake an out-of-reach object has nothing whatsoever to do with the one that makes it possible to use a hammer-like stone to break the hard shell of a fruit?

Much for the same reasons, a structural theorist must face the related problem that data vastly underdetermine theory, that is to say, that quite a number of possible models are compatible with the same set of data.

Holding a structural view, Kohler was well aware of the arbitrariness problem (what is intelligence? how is it delimited from other capacities?), and it was in connection to this problem that he envisaged a fundamental role of comparison. This, in fact, is his second declared "set of interests" in pursuing the study of intelligence in apes:

> The second aim is theoretical—writes Kohler in the same introduction quoted above—Even assuming that the anthropoid ape behaves intelligently in the

sense in which the word is applied to man, there is yet from the very start no doubt that he remains in this respect far behind man, becoming perplexed and making mistakes in relatively simple situations; but it is precisely for this reason that we may, under the simplest condition, *gain knowledge of the nature of intelligent acts*. (Kohler, 1976, p. 1, italics added)

Patterns of similarity and difference among closely related species might offer powerful cues as to the constituency of the underlying mechanisms that can generate them. On the one side, they show natural groupings that contribute to answer the question of what goes with what and, hence, offer a non-arbitrary basis to delimit boundaries of domains.

On the other side, structures responsible for species differences must be so modeled as to allow their being interrelated by plausible evolutionary transformations (where "plausible" means that take into account parameters like the evolutionary distance between the species compared, directional trends across a number of related species, functional viability at all intermediate stages, etc.). In view of the fact that one of the defining characteristics of structures is their being internally highly interrelated, this is no trivial contraint: tight interrelation means that even a slight local modification is likely to reverberate diffusely and snowball into widespread reorganizations. Hence, choice among possible models might be further reduced by the requirement that they must be capable of supporting a given transformation. Within such a perspective, a double role for the comparative study is thus provided. Structures and the transformations relating them might enter in a beneficial mutual relation: comparative modeling of the first leads to reconstructing the evolutionary transformation, which in turn feeds back in constraining and hence refining the hypothesized models. Far from being vicious, this circle can generate successive convergences that substantially reduce model underdeterminacy. It is perhaps no accident that the initial rebirth of the original intent of comparative psychology took place with the study of one of the most highly structured, theory dependent, cognitive domain, language capacity, in man's closest relatives, where the working of this process could be amply exploited (see Gardner & Gardner, 1971; Premack, 1976; Rumbaugh, 1977; Terrace, Petitto, Sanders & Bever, 1979).

Exactly the same set of reasons motivates the addition of another fundamental dimension to the study of cognitive capacities: that of their ontogenetic development. As shown in embryogenesis, at all levels, ontogenesis of structures proceeds by progressive differentiation, on the one side, and increasing integration, on the other side. Corresponding structures can then be compared (much as in evolutionary comparison) at different stages of this process. The interplay between structure and developmental transformation can then generate the same positive effects on constraining models as in the case of phyletic transformation. On the one hand, structures can be investigated before or at a partial level of integration. By seeing developmentally related groups of phenomena, we can again understand more easily what goes with what, what is more likely to depend on a unitary underlying mechanism. On the other hand, and most importantly, the way adult structures are organized might depend to a large extent by the very way they develop (see, for example, Chomsky's argument [1975] that a theory of language acquisition is *ipso facto* an explanatory theory of language capacity). The theory of cognitive capacities and functioning developed by Piaget, which will be illustrated in the next chapter, and the evidence supporting it provide ample demonstration of this claim.

Finally, evolutionary and ontogenetic transformations might be related in a much more fundamental way than in the similar roles they play in theory construction by constraining models of capacities. It has been repeatedly argued in the past (de Beer, 1930; Garstang, 1922) that ontogeny is the main locus of evolutionary change. With recent critiques to the "modern synthesis" paradigm of evolutionary theory (Eldredge & Gould, 1972; Stanley, 1979), this position is currently the object of renewed theoretical interest (Gould, 1977; Wake & Larson, 1987; Gottlieb, 1987). Recent techniques in dating evolutionary relatedness ("molecular" clocks), dominant stability in stratigraphic records ("stasis"), as well as other converging pieces of evidence, put more and more in question the slow and gradualistic view of evolutionary transformation (by piecemeal accumulation of small changes) originally maintained by Darwin and strongly reaffirmed in the modern synthesis. The tempo of macroevolution seems to be much more rapid: to take a concrete example, evolutionary distance between man and chimpanzee as measured by similarity in the genetic make-up appears to be much smaller than previously hypothesized and to translate into a few million year divergence, not many more than those separating modern man from its first australopithecine ancestor (see Nute & Mills, 1986). Yet the macroscopic differences of these two species in many structures are wide. This paradox means that small changes at the genetic level must somehow translate into large differences. The combined effects of structural interrelation and developmental transformation might provide just such a mechanism. In fact, small changes taking place early in ontogeny, that is, in "parent" structures that will later differentiate, will be projected onto all "descendant" structures and hence widely amplified. The final effect, as seen in the adult, is, therefore, much more dependent on the time of occurrence of such modification (the earlier, the greater), than on its magnitude (see chap. 7, for a possible example of this kind). Haeckel's (1866) traditional and influential argument that "Phylogeny is the mechanical cause of ontogeny", giving rise to the celebrated doctrine of ontogenetic recapitulation, is thus reversed into its opposite.

For these reasons, consideration of ontogenetic development, joined to a structural standing, will be the major focus of this book (as evidenced in its title), in the belief that, even more than simple comparison, the comparative study of the cognitive ontogenies of man's related species might provide an essential key to the explanation of human cognitive capacities.

2 The Theoretical Framework

Francesco Antinucci
Istituto di Psicologia, C.N.R.
Rome, Italy

The approach to the study of cognition that will be followed in this book is grounded in the epistemological and theoretical work of Jean Piaget. Beside the vastness and empirical value of his specific findings in a half century of uninterrupted search for the origin and structure of human cognition, the general principles governing cognitive organization Piaget developed and substantiated along this long work represent an ideal tool to pursue the study of cognition in terms of the biological perspective outlined above.

In the introduction to his studies on the origins of intelligence, under the title "The biological problem of intelligence", Piaget (1974, p. 3–5) writes:

> Intelligence is an adaptation. . . . Certain biologists define adaptation simply as preservation and survival, that is to say, the equilibrium between the organism and the environment. But the concept loses all interest because it becomes confused with life itself. . . . There is adaptation when the organism is transformed by the environment and when this variation results in an increase in the interchanges between the environment and itself which are favorable to its preservation. . . . To say that intelligence is a particular instance of biological adaptation is thus to suppose that it is essentially an organization and that its function is to structure the universe just as the organism structures its immediate environment.

Functional continuity and structural discontinuity are the cornerstones of this view.

Adaptation realized by cognition, in the sense of the transformative process defined above, is the product of functional processes that are invariant across "organisms" (whether these are conceived as different species, or ontogenetic phases of an individual). Their action, however, generates and successively transforms structures that, as organized and organizing totalities, are discontinuous with each other, i.e., that are not merely quantitative amplification of each other. Continuity of function provides the indispensable common basis on which comparison can be effectively operated, while the possibility of discontinuity in the structures produced allows for specificity at each level. In this way, the comparison is not trivialized as in behaviorism, where functional continuity is instead accompanied by structural continuity (see the "atomistic" assumption, discussed in the preceding chapter).

Let's now survey these fundamental functional constants, and briefly exemplify the structures they generate. The first of them is the *construction of invariants*: cognition attempts, at all levels, to construct invariants. Contact with the external world generates at all instants an ever-changing flux of effects in the organism: no two events are identical. Apprehension of this infinite variety in itself would be useless. Given organs that can register whatever physical dimension of the external world, i.e., that alter their state as a consequence of its variations in a sufficiently permanent way, the specific activity of cognition, as opposed to other adaptive mechanisms that also react to such variations (like, for example, skin-darkening by ultraviolet radiations), is to relate such registrations by establishing correspondence transformations between or among them.

At its elementary levels, this process leads to well-known products: from the familiar categorical reunion of sensory stimuli to the construction of perceptual constancies, thereby different sensorial images (as resulting, for example, from an object in movement, or seen from a different angle or at different distances) are perceived as invariant. Less obvious, because of their structural discontinuity, are instead its innumerable higher level products that Piaget investigated all the way to their uppermost term, the construction of scientific theories.

Still at the sensorimotor level, one can trace the construction of the object-concept (what is commonly called "object-permanence"): under the appropriate transformative conditions, "objects" appearing to the subject at different times and places are unified as "the same object".

Object permanence is the last perceptual and the first conceptual invariant: objects are unified across the spatial and temporal discontinuities of their perceptual appearance, hence they are constructed as independent from the subject's own perceptions (Piaget, 1971a).

At this level, however, the physical appearance of the object, even if discontinuous, stays the same. Higher level invariants are those that, instead, enable to construct an object as "the same" when its form is altered. These are the well-known "conservations" (of substance, weight and volume): objects are unified across transformations of their physical form (Piaget & Inhelder, 1962).

Finally, objects can be construed as invariants no matter what their physical appearance: they are unified across all possible forms. In this last case, they are "the same" *qua* "units".

Construction of invariants guides also the structuring of relations among objects. For example, in space, where they allow to keep constant the order relations among objects independently of the subject's displacements relative to them. They range from invariance preserved through transformations of perspective to invariance preserved through map projections (Piaget & Inhelder, 1967).

Invariants are also constructed across transformations of spatial and order relations among objects. When they include also the identity of the objects involved (i.e., the "unit" invariant), the fundamental invariant of numerical quantity is constructed. Sets of objects are "the same", no matter what the spatial disposition and identity of their elements, if they "number" the same (Piaget & Szeminska, 1941).

At all levels, invariant construction generates as its complementary process that of giving order to the differences encompassed. Structures produced are, thus, capable of not only establishing equivalences, but also ordering non-equivalences (as in constructing ordered series). In the example just quoted, sets of objects can be not only numerically equated, but also precisely ordinated according to their disequivalence.

Finally, at its highest level, cognition engages in constructing invariants across invariants themselves (or across invariants of invariants, and recursively on; that is, invariants of all possible orders). Cognition engages in constructing "theories". Though the structural discontinuity among theories, the products of this level, may be immense (as, for example, between naive and formal theories of object motion) the process is still continuous. Any textbook on the history of science will evidence the novelty constituted by the introduction of crucial "new" invariants, like the principle of inertia (unifying stasis and movement) or the particle theory of atoms (atoms are the same under their constituent parts). One of the most transparent example of this process is Darwin's evolutionary theory. Different species, each of which is already a (first-order) invariant across different individuals, are unified under the appropriate transformation: that operated by natural selection. A higher order invariant across species is thus constructed. But Darwin doesn't stop here. Besides reducing cross-species difference, he attempted to reduce to invariance also within-species differences (obviously, not individual variation, but systematic variation, i.e., variation already construed as an invariant, like differences of variety or sex). This is accomplished by constructing another invariant: within species, differences are unified under the transformations operated by sexual selection.

In fact, the two most famous books by Darwin, *The origin of species* and *The descent of man*, are not so much separated because of the empirical topics dealt with (man vs. other species), but rather by their being separately devoted to the construction of each of these two different invariants.

The fundamental continuity of the functional process, extending from sensorimotor to theory construction, joined to the wide discontinuity of the structures thus produced raises the problem of the relationship among these structures, and especially, of the passage between one and the other. It is this inescapable question that led Piaget to investigate (and dedicate to it a major part of his effort), on the one side, the relation between science and cognition, in what he termed "genetic epistemology" (Piaget, 1971b, 1972) and, on the other side, the mechanisms of passage, or what he called the "theory of equilibration" (Piaget, 1977).

A second, fundamental functional constant of the cognitive process is its being based, again at all levels, on the action of the cognizing subject.

> Human knowledge is essentially active. To know is to assimilate reality into a system of transformations, to know is to transform reality in order to understand how a certain state is brought about. By virtue of this point of view, I find myself opposed to the view of knowledge as a copy, a passive copy, of reality. . . . To my way of thinking, knowing an object does not mean copying it—it means acting upon it. It means constructing systems of transformations that can be carried out on or with this object. (Piaget, 1971b, p. 15)

At the most elementary level, this is the concrete, physical action of the subject on the external world. Actions that organize in action schemata give rise to the whole of what is called "sensorimotor intelligence". At higher levels, action might become internalized, that is, it can be performed mentally rather than physically. By becoming mental, actions acquire much more power than their physical counterpart. They can be executed much faster and, especially, they can be executed in ways that are not always possible to physical actions: for example they can be reversed or executed simultaneously. They can furthermore be carried out on mental objects, such as "symbols", that stand for any kind of external object, or "abstract" objects, such as the invariants considered above.

Obviously, this last kind of action generates structures that are incomparably wider in scope, richer and more powerful than those constructed in sensorimotor intelligence. Construction of the "conservation" invariants would, for example, be impossible without reversibility. Yet they remain actions: this functional continuity implies that, even if vastly discontinuous with sensorimotor ones, these structures will bear in their organization the inevitable marks of their origin.

This directly leads us to another important constant, one which is grounded, in fact, in the nature of the cognizing actions.

> There are two possibilities. The first is that, when we act upon an object, our knowledge is derived from the object itself. . . . But there is a second possibility: when we are acting upon an object, we can also take into account the action itself. . . . A child, for instance can heft objects in his hands and realize that they have different weights—that usually big things weigh more than little ones, but that sometimes little things weigh more than big ones. All this he finds out experientially, and his knowledge is abstracted from the objects themselves. But I should like to give an example just as primitive as that one, in which knowledge is abstracted from actions, from the coordination of actions, and not from objects. This example, one we have studied quite thoroughly with many children, was first suggested to me by a mathematician friend. . . . When he was a small child, he was counting pebbles one day; he lined them up in a row, counted them from left to right, and got ten. Then, just for fun, he counted them from right to left to see what number he would get, and was astonished that he got ten again. He put the pebbles in a circle and counted them, and once again there were ten. He went around the circle in the other way and got ten again. And no matter how he put the pebbles down, when he counted them, the number came to ten. He discovered what is known in mathematics as commutativity, that is, the sum is independent of the order. But how did he discover this? Is this commutativity a property of the pebbles? It is true that the pebbles, as it were, let him arrange them in various ways; he could not have done the same thing with drops of water. So in this sense there was a physical aspect to his knowledge. But the order was not in the pebbles; it was he, the subject, who put the pebbles in a line and then in a circle. Moreover the sum was not in the pebbles themselves; it was he who united them. The knowledge that this future mathematician discovered that day was drawn, then, not from the physical properties of the pebbles, but from the actions he carried out on the pebbles. This knowledge is what I call logical mathematical knowledge and not physical knowledge. (Piaget, 1971a, pp.16–17)

On the one side, different objects react in different ways to the same action and different actions bring about different modifications of the state of an object. Through this process, properties and behaviors of objects are determined. On the other side, the temporal and spatial organization of actions themselves can introduce a related organization in the objects acted upon that has nothing to do with their physical characteristics or behaviors. In this case, it is not the specific type of action (and not the specific type of object) that is relevant, but the way actions are organized with one another. At all levels, therefore, we will be able to distinguish between a "logical" component and a "physical" component of cognitive structures, and, as we shall see below, this distinction will play a fundamental role in comparing human vs. nonhuman primate cognitive organization and development.

To sum up, comparison of structures across species, as well as developmental phases, will enable us to characterize the specificity of each cognitive organization and, hopefully, determine in what exactly their difference lies. They will be the "meters", so to speak, of our comparison.

If we can then trace how these differences are connected to the operation of the functional constants, we will be able to suggest the modes (and, possibly, the causes) of their differentiation.

We will not consider at all, instead, the problem of the innate vs. acquired origin, or of the maturational unfolding vs. constructed development of the cognitive organizations found, though we are well aware of the strong and explicit position taken by Piaget on this topic. In the wider perspective taken here, whereby transformational processes relating cognitive organizations are seen as spanning not only over developmental phases of the same individual but also across species, this problem loses most of its appeal. In fact, interest in these two opposing positions derives not so much from an empirical curiosity about the mechanism of development *per se*, but rather from the fact they are (more or less explicitly) tied to very different conceptions of the nature and structure of the resulting organization. In one case, this tends to be seen as essentially homogeneous at all levels, flexible, and, especially, step-by-step modifiable in almost every part; in the other case, as discontinuous in its levels, specific in its part structure, and limitedly modifiable except through reorganization. In other words, it is the theory of cognitive organization that is really argued upon.

In the view put forth here, cognitive organization is structured by its constructive processes which are invariant at the level of the functional constants just described. In this sense, a set of "possible" cognitive organizations (albeit an infinite one, with respect to what is effectively realized) is, at least in principle, defined: a structure is possible if it is a transformation of (or, equivalently, it is constructed from) another structure by means of the "allowed" functional processes, and so on recursively.

Within this view, therefore, intrinsic constraints on cognitive organizations do not depend on where the transformative process takes place: whether within one and the same individual organism, or across individual organisms, or across species of organisms. Obviously, for the transformation to be possible, the departing structure to transform must be "available" to the organism effecting it (this is a necessary but by no means sufficient condition), but it is irrelevant whether he himself has previously constructed it or it has been handed to him by an historical or genetic "memory."

Of course, we do not wish to deny or minimize the importance of these differences, their consequences and their study, but only claim that they are not central to the understanding of how cognitive organization is intrinsically structured. In fact, it is our impression that these topics can be studied in a much more effective way when a great deal more will be known about the last.

A final *caveat* may be necessary: the view just exposed should not be construed as implying a necessary identity between ontogenetic and phylogenetic processes. The two are obviously related, since they share the same constraints on possible structures and transformations relating them, but this in no way implies the much stronger claim that the same structures in the same succession will be built by the two processes.

II LONGITUDINAL STUDY OF COGNITIVE ONTOGENY: BIRTH TO STAGE 4

3 Early Sensorimotor Development in Gorilla

Giovanna Spinozzi and Francesco Natale
Istituto di Psicologia, C.N.R.
Rome, Italy

INTRODUCTION

The highland and lowland rain forests of Central and West Africa, from Cameroon to Uganda, are the natural habitat of the largest of the extant primates, the gorilla.

With other living apes, this genus is taxonomically closely related to the human species, sharing with it the same superfamily Hominoidea, and differing from all other species of Old World monkeys, which belong to the superfamily of Cercopithecoidea.

Traditionally (e.g. Napier & Napier, 1967), the Hominoidea have been divided into three families: Hylobatidae or "lesser apes" (gibbon and siamang) living in Asia, Pongidae or "great apes", including orangutan, gorilla and chimpanzee and Hominidae containing our own species.

Although it is still agreed that all great apes and man shared a common ancestor when lesser apes had already branched off, there is no universal agreement on how they are internally related to each other. Beginning from the sixties, traditional morphological classification and branching dates based on fossil evidence have been repeatedly called into question by the results of relationship and dating studies based on macromolecular similarity and evolution (for a review, see Nute & Mills, 1986). Rather than following this long and complicated debate, it seems useful to present where a majority consensus appears to stand today, at least for those points that are agreed upon (see Pilbeam, 1984; Schwartz, 1986; Groves, 1986).

According to this view, apes and man shared for a long period a common ancestor that made its first appearance between the late Oligocene and the

early Miocene (30 to 20 million years ago), emerging from a basic stock
of Old World monkeys. It seems almost certain that the gibbons diverged
from the larger hominoids probably in the earlier Miocene. Then, by middle
Miocene (around 16 m.y. ago), the divergence of the Asian branch occurred, giving rise to ancestors of present day orangutan (Pongo). The
African branch of Miocene hominoids is without fossil representation;
nevertheless molecular evidence suggest that gorilla, chimpanzee and hominids split between 10 and 6 m.y. ago. Thus man appears to be more closely
related to the African apes than to the Asian one. The internal relation
of Pan, Gorilla and Homo is the most debated point: more traditional
orientations (e.g. Schwartz, 1986) have chimpanzee and gorilla classed
together into a Panidae family, and man in the Hominidae. Macromolecular studies and recent morphological comparisons (e.g. Groves, 1986)
consider instead Pan as the sister group of Homo, and Gorilla an earlier
branching off. According to this view, then, chimpanzee is more closely
related to man than gorilla is, and, possibly, more closely related to man
than to gorilla.

Three subspecies are recognized within the genus Gorilla: the eastern
lowland gorilla (*Gorilla gorilla graueri*) which is found in forests of eastern
Zaire; the eastern highland gorilla (*Gorilla gorilla beringeri*), the typical
mountain gorilla, which inhabits Virunga extinct volcanoes; the western
lowland gorilla (*Gorilla gorilla gorilla*), which is the subject of the present
study, inhabiting mountain and riverine forests stretching between western
Cameroon and Congo river, from sea level to 700 m. altitude.

Gorillas spend much of their day searching for and eating food. Their
diet is mainly vegetarian, consisting of stems, leaves, shoots and fruits.
They are predominately terrestrial feeders but their foraging behavior varies greatly according to where they live. In areas that are comparatively
richer in trees, they have been found to exhibit arboreal feeding habits
(Goodall, 1974). Western gorillas live in small social groups, numbering
from 5 to 13 individuals.

A typical gorilla group contains one "silverback male", the oldest male
whose short fur on his back and flanks has become gray, one or more
blackback males, several females, juveniles and infants. The silverbacks
are responsible for group cohesiveness, constituting the focus for all females
and infants, which spend a large amount of time in their proximity during
day-resting; when the group moves in forest, the silverback male walks at
its head emitting from time to time typical vocalizations to coordinate group
movement (Fossey, 1972).

Of the three subspecies, the western gorilla, is the smallest, with adult
males averaging 153 kg. and adult females about half the weight of males.
The cranial capacity is in the range of 420–752 cc. in adult males and 340–
595 cc. in adult females, which represents 0.3% and 0.5%, respectively,

of body weight. Like Old World monkeys and all apes, gorillas have the same dental formula of man, but unlike humans, the canines are larger in adult males than in females, whereas molar teeth are very large in both sexes.

The males begin to be sexually mature at about 8 years of age and are fully mature at about 11 when their back turns silver. The females reach sexual maturity between 8 and 9 years. The length of their menstrual cycle and gestation period is very similar to that of the human females: the first is about four weeks and the second averages 263 days (Napier & Napier, 1967; Fossey, 1979).

As for maturational characteristics, infant gorillas complete their deciduous dentition during the second year of age: the first teeth to appear are the incisors, followed by molars, premolars and, finally, by canines. (Schultz, 1968). Permanent dentition is complete only at 10–11 years.

The average weight of infant gorillas is about 2.1 kg. at birth and the cranial capacity is about 400 cubic centimeters at 10 weeks of age. But while weight strongly increases in adulthood, brain growth from infancy to adulthood is very small. In fact, as we saw before, the cranial capacity is respectively only 0.3% and 0.5% of body weight in adult males and females. These data are very different from those of humans whose cranial capacity is 2.1% of body weight at adulthood. Consequently, the gorilla's brain is more developed at birth and is smaller in relation to adult body weight than humans'.

The infant gorillas form long and stable associations with their peers and siblings and have close relations with their mother extending even into adulthood.

Fossey (1979) observed in the field and described the behavioral development of mountain gorilla from birth to 36 months of age. For their first month, infants are always carried in ventral position, even when they begin to be able to cling on the maternal ventral fur. From 4 months of age they begin to move away from their mothers, but always remain within arm's reach. The first social interactions with other young animals appear during this period, but self-play is most frequent. At one year the infant gorilla is still suckling, even if the mother often prevents it. During the second year of age the gorillas travel in dorsal position only for long distances, but more often follow their mothers independently. Although suckling continues, they begin to strip leaves as a mean to obtain food. Social play increases and social responses toward adults and attraction toward the silverback male begin to be exhibited. During the third year independent locomotion increases even if the infants are still carried sometimes by their mothers. Suckling is still observed but leaves and vines are most often eaten. Social play and interactions with the other members of the group increase and more time is spent in proximity of silverback male.

As it can be seen, many physiological characteristics, such as length of gestation period and menstrual cycling, length of physical maturation period are close to that of man, as are the prolonged infant phase and the long dependence of infants from mother's care. In fact, these features are closer to man that to other monkey species.

The present study maps the sensorimotor intelligence development of an infant female gorilla (*G.gorilla gorilla*) from birth to 15 months of age and attempts to compare it to that of the human child, as described by Piaget (1974).

Studies of the cognitive abilities of gorillas have been and still are quite rare (as opposed to many other monkey species, or to chimpanzees). The majority of them are, furthermore, framed in terms of comparison with the more studied and more known ape, the chimpanzee.

The first of these studies and, still today, one of the most extensive is that of a young female gorilla, Congo, by Yerkes (1927a, 1927b). Yerkes had in mind the classical studies of Kohler (1976) on the intellectual abilities of chimpanzees, and many of his experiments were conducted within that frame. Some of his interesting results will be discussed below, but in general Yerkes found her performance inferior to that of the chimpanzee, especially in tasks involving manipulative abilities. Knobloch and Pasaminick (1959) tested cognitive development of a gorilla infant in her first two years of life using the Gesell Developmental Examination (Gesell & Amatruda, 1951). The authors compared the gorilla's development with that of the chimpanzee and the human infant, and found that the gorilla developed faster than the other two species, but reached a much lower level of conceptualization and abstraction capacities. Knobloch and Pasaminick related the progressively more advanced capacities found in the three species to their progressively longer maturational development.

Rumbaugh and McCormick (1967) found no difference with chimpanzee or orangutan in the performance on oddity concept problems. Parker (1969) compared the manipulative abilities of apes and found that the gorilla was less responsive to object manipulations than either chimpanzee or orangutan.

Two recent studies (Redshaw, 1978; Chevalier-Skolnikoff, 1977) have attempted to investigate sensorimotor intelligence development in gorilla infants within a piagetian framework. Their general results (with one apparent exception, which will be discussed below) are in good agreement with those of the present study. Both studies, however, raise some methodological problems. Redshaw tested sensorimotor development exclusively by means of the Uzgiris and Hunt (1975) scales of infant development. These scales have been developed and standardized on populations of human infants and their items have been selected, through a delicate and long trial process, in order to minimize the testing time and effort, by searching for only those behavior patterns that turned out to be more sensitive and reliable in scaling. There is no guarantee that these items will play the same function when employed with species whose maturational and physical characteristics are different from those of man. The original piagetian "clinical" methodology was a discovery procedure for cognitive structures and functioning, while the scales are simply a tool of assessment of conformity to already known developmental norms. It seems to us that in trying to map the unknown course of development of nonhuman primate species, it is essential to use the first of these methods.

Chevalier-Skolnikoff used a combination of free observation and testing according to the Uzgiris and Hunt scales. She adds, however, that the testing procedure was modified, but does not specify how. Furthermore, she doesn't report any result, as Redshaw does, in terms of the scales items. She devised a peculiar classification of behaviors according to sensory modality and body part (not to be found anywhere else in the literature on sensorimotor intelligence) whose theoretical relevance is unjustified.

These problems make it difficult a detailed analysis of the results of these studies and a comparison with our. Some of the points raised will, however, be dealt with in the discussion.

SUBJECT AND METHOD

The gorilla examined in this study is an infant female (*G.gorilla gorilla*), Romina (Rm), born in the Rome Zoo on April 18, 1980. She was reared by her mother for the first month and a half, at which time her mother began to refuse to hold and feed her. She was then separated from her and hand-reared by a caretaker.

Data for the first month and a half of her development could, therefore, be collected only by observing her spontaneous behaviors from outside the cage where they were living. This was done during hour-long sessions that occurred once a week at different times of the day.

Following her abandonment, it became possible for the experimenters to interact with the infant. Data were then collected through the standard "clinical" methodology of piagetian longitudinal studies: the subject was both observed while spontaneously playing with objects offered to her by the experimenters and given specific test-situations to probe her abilities. Objects and situations chosen varied according to her level of development.

Testing continued to occur in hour-long sessions once a week, until Romina was three months old. From three to fifteen months, sessions took place, instead, once every three weeks. All sessions were videotaped and then transcribed. Original 3/4" black-and-white cassettes are available.

RESULTS

The analysis of data tried to identify in our subject the first four stages of sensorimotor intelligence development, as described by Piaget (1971a, 1974) for the human infant, and characterize them by comparison to those of the human infant. To this purpose, presence or absence of all the behavioral indexes that are typical of each stage was noted together with the date of their appearance. Tables 1-5 below summarize some of the items that are crucial for the identification of each stage, as well as for the developmental sequence across stages. The items selected were, furthermore, those that proved to be among the most stable and reliable to assess across the four species of nonhuman primates of our longitudinal studies, and could thus provide a more effective basis for comparison both within nonhuman primates and with human children. The tables report the age (in days) of first appearance of the behavioral item in question, except for the last item of Table 5 ("Typical reaction") where the age is that of the last testing session where the behavior was still observed.

Stage 1

In stage 1, the cognitive activities of the human infant consist, fundamentally, in releasing and practicing the initial sensorimotor reflexes centered on vision, suction and prehension.

Through this exercise, in the course of the first month of life, more and more of the external stimuli get assimilated, while, at the same time, the original reflexes adapt to them and differentiate, giving rise to "primary" sensorimotor schemata, that is, schemata where "objects" are directly assimilated by the subject's organs. Such schemata represent the fundamental way of knowledge of this stage: external objects are known insofar as they are mouthed or sucked, fixated or followed by the eye, tactually and kinesthetically felt by the hand. It is important to note that throughout this stage these schemata stay separate from one another: there is a world of visual images, one of oral images and one of hand images and they are unconnected to one another. The object that is seen is not yet the same object that is touched and not yet the same object that is mouthed. Co-ordination and unification of these separate images mark, in fact, the beginning of the next stage.

The development of the infant gorilla in this stage was virtually identical to that of the human infant. It displayed the same behavioral repertoire and at about the same ages as the human child. Three sets of activities, each centered on one of the main organs, the eye, the mouth and the hand, can be distinguished. With respect to vision, for the first two weeks, our subject was incapable of fixating even a stationary stimulus (like the face

TABLE 1.
Stage 1 of Sensorimotor development in Gorilla

	Age in days Rm
VISION	
- Partial visual pursuit	38
SUCKING	
- Orienting to contact within mouth-area	32

of its mother, a light, or a brightly colored object) for more than few instants. Subsequently, visual contact with such stimuli became sustained, but it was immediately lost as soon as the stimulus began to move slowly in front of her eyes. Finally, at the beginning of the second month, the gorilla became capable of visually following an object slowly moving in front of her eyes on an arc-like trajectory, by slightly turning her head in the right direction; but only for a short stretch, then contact was again lost ("Partial visual pursuit" in Table 1).

Both the sucking and the rooting reflexes were well developed during the first two weeks of age, but with very little accommodation and orientation to stimuli. For instance, while in ventral contact with her mother, as soon as her mouth came into contact with the mother's fur or the aureola of the nipple, she began to tap repeatedly the surface with circular movements of the head, while attempting to suck. There was no accommodation: she tried to suck everywhere and her movements were not oriented toward the nipple.

At the end of the first month accommodation could, instead, be noticed: when any part of the infant's lips came in contact with the nipple or its surrounding area, she immediately opened the mouth and turned her head in the right direction, centering the nipple without any search or attempt ("Orienting to contact within mouth-area" in Table 1).

Because of our conditions of observation, we could not distinguish clearly the further progression of this adaptation, i.e., when even contact outside of the mouth area (cheeks or chin) gives rise to a correct orientation of the mouth toward the nipple, as observed in children and in the other species tested. Also the grasping reflex was as in children. The simple contact of the palm of the hand with the mother's fur caused the hand to close and grasp. Prehension was, however, soon released and eventually renewed, thus giving raise to the typical closing and opening movements of the hand observed in this stage when the palm is tactually stimulated by some object.

For the whole first month of life, the gorilla was unable to cling to her mother's fur when not supported, and had to be constantly carried by her.

As in children, movements of the arms were impulsive and undirected: when the hands lost contact with the mother's body, the arms waved in the air.

Motor abilities related to the trunk, limbs and head were also very similar to those showed by children in the first month of life. Contrary, as we shall see, to the other nonhuman primate species tested, none of the items characterizing the development of independent locomotion could be scored during this whole period. The gorilla could not even support her head erect and her motor development appeared as slow as that of children.

Stage 2

The adaptation and differentiation of the primary schemata of vision, suction and prehension, already seen in the preceding stage, continues to increase during this stage. But, most of all, this stage is characterized by the development of their intercoordination. When the hand can grasp what is held in the mouth or bring to the mouth what it has grasped, and when the hand can grasp an object that the eyes see, or carry in front of the eyes an object that the mouth is holding, then the hitherto separate images that the eye, mouth and hand have of the object unite into one, and objects begin to exist as such. Children undergo this development from the second to the fourth month of age.

A clear differentiation of the sucking schema, and the same one that is seen in children at about the same age, emerged in our subject at the beginning of the second month. From her 38th day the gorilla started to show a repeated protrusion of the tongue when her mouth accidentally touched her hand ("Tongue protrusion" in Table 2). While in the preceding days such a contact activated sucking, now the schema has become differentiated into a non-functional vs. a functional sucking. As a result, objects are now primitively categorized: of all "suckable objects" some are for feeding and some not, and they are now recognized as such.

In the same day, she manifested the first (unsuccessful) attempt to hand-mouth coordination: in fact, after an accidental contact of her hand with the lips, she tried to grasp the finger with her mouth. As soon as contact with the hand was lost, she tried to establish it again by large movements of the hand vaguely oriented toward the mouth. Coordination was, however, still lacking and no success ensued.

At the same age, vision was still no more adjusted that we saw in the preceding stage. She was visually attracted by and tried to follow her own hand movements, which, as we saw, were still largely uncoordinated, but she lost contact fairly soon. It was only four weeks later that she became capable of following systematically slowly moving objects or the displacements of her mother through a whole 180 degree arc, both forward and

TABLE 2.
Stage 2 of Sensorimotor development in Gorilla

	Age in days Rm
ADAPTATIONS OF SUCKING	
- Tongue protrusion	38
ADAPTATIONS OF VISION	
- Complete visual pursuit	66
ADAPTATIONS OF PREHENSION	
- Holding an object	66
COORDINATIONS	
- Thumb sucking	56
- Bring to mouth grasped objects	66
- First unsuccessful attempts to visually guided hand prehension	66
- "Buccal prehension"	88
- First successful attempts to visually guided hand prehension	110
- Fully developed visually guided hand prehension	168

backward, without losing contact ("Complete visual pursuit" in Table 2). At this age, the gorilla also began to be able to grasp and hold an object in hand for a substantial period of time, without immediately releasing her grasp ("Holding an object" in Table 2). Both the progression and the timing of these developments were practically indistinguishable from those of children. The development of schemata intercoordinations, on the other hand, showed some interesting differences.

The gorilla began, as children do, with the simplest coordination, that between the hand and the mouth, manifested in the ability to bring to the mouth the hand in order to suck the thumb ("Thumb sucking" in Table 2). At this time any coordination between grasping and vision was lacking: she made no attempt to grasp objects that she saw in front of her eyes. The second developmental step was again as in children: coordination between prehension and mouthing became established. She was able to bring a grasped object to the mouth, and to grasp an object hold in mouth. At the same time the first attempts to grasp objects at sight began. They were, however, all unsuccessful: the hand was projected toward the object, but failed both in direction and in depth adjustments ("First unsuccessful attempts to visually guided hand prehension" in Table 2).

At the end of the third month the gorilla became capable of displacing her body in space (see Table 3) and as soon as this happened she developed a peculiar coordination that is never seen in children: she began to move toward the objects she saw and grasp them with her mouth. We labeled this pattern "buccal prehension". The reason of this development appears to be the following: the coordination of vision and locomotion is simpler to achieve, and hence might occur earlier, than that between vision and hand-prehension. Since eyes and mouth occupy fixed position with respect to each other on the head, any movement of the eyes toward the object will automatically target the mouth onto the object. On the other hand, the hand moves through schemata that are independent of the eyes and the head movements, hence the coordination of visual and hand-movement schemata is more difficult to establish (and, in fact, it is the last one to develop in children, and, as we shall see, also in all our species).

Since grasping at sight continued to be unsuccessful for quite a time (first successful attempts appearing toward the end of the fourth month), buccal prehension became during this period the dominant pattern of interaction with objects. Sometimes objects were first grasped with the mouth and then taken from the mouth with the hand, by means of the already established hand-mouth coordination.

Buccal prehension and its related behavioral patterns are, obviously, never seen in the development of the human infant: contrary to the gorilla, the child has no capacity for independent locomotion during this period and will not have one for a long time.

Visually guided hand-prehension developed, instead, through the usual two steps found in children. From 110 days attempts to grasp at sight became for the first time successful. The hand still made large and wrong movements toward the object, but now, when the hand entered into the visual field and was perceived together with the object, it was guided by vision and its trajectory was slowly adjusted toward the object (as if it were an independent object) ("First successful attempts to visually guided hand prehension" in Table 2).

From this moment on, mouth use in prehension progressively decreased in favor of hand use, which got better and better. Mouth use was still favored when grasping objects presented some difficulties, either owing to the size of the object (objects that were very small), or to the situation in which they were presented (objects that were attached to a string or objects partially hidden by a screen). The behavior shown by our gorilla in this context is identical to that found by Knobloch and Pasaminick (1959) in their subject:

> When Colo became more interested in small things, she was very inept at securing them, and she began to develop the use of her mouth as a third prehensile organ. This pattern of mouth reaching she elaborated, and in many instances seemed to prefer it, for example, as a method of removing the cube from the cup or of securing the pellet. (Knobloch & Pasaminick, 1959, p. 702)

Finally, at the beginning of the fifth month, hand prehension completed its development: the infant became able to grasp any object at sight with

TABLE 3.
Development of Locomotion in Gorilla

	Age in days Rm
Phase 1	
- Supporting head erect	66
- Rolling to prone	88
Phase 2	
- Sitting with forearm support	88
- Standing on the fours	88
- Walking with forearm support	88
Phase 3	
- Walking on all fours	99
- Sitting without support	125

a single, precise movement of the hand ("Fully developed visually guided hand prehension" in Table 2). Prehension by hand became from then on the dominant pattern, completely reversing the preceding situation with respect to mouth prehension.

Mouth prehension, however, did not disappear entirely, even after perfect coordination between hand and vision had been achieved. Significantly, it remained as a more primitive, simpler pattern on which our subject fell back when her attempt to grasp at sight got, for whatever reason, frustrated.

Finally, as seen above, locomotion development begins and is completed in this stage (Table 3).

Stage 3

Once the originally separate schemata have been fully integrated, so that to one and the same object are now simultaneously attributed visual, oral and tactual-kinesthetic properties, schemata begin to be organized into structured sequences, from which properties of objects that go beyond the primary ones can be derived. Perhaps the most important of these initial structures, the ones that according to Piaget are most typical of the third stage of sensorimotor intelligence, are "secondary circular reactions". The typical behavior pattern that manifests a secondary circular reaction is the following: the subject acts on an object and as a consequence the object produces some effect that the subject notices and immediately tries to reproduce by repeating the action on the object. For example, the infant shakes a rattle that produces sound, he then tries to reproduce this effect by shaking again the rattle and so on. These sequences represent the initial exploration of objective cause and effect relations. In fact, while every

TABLE 4.
Stage 3 of Sensorimotor development in Gorilla

	Age in days Rm
Sequences of primary schemata applied to objects	195
Secondary circular reactions	—
OBJECT CONCEPT	
- Visual accommodation to rapidly moving objects	168
- Recovering partially hidden objects	180
- Searching for hidden objects at the point of disappearance	195

action of the subject "causes" an "effect," in simple assimilating actions the cause remains internal to the subject (in his efforts or movements), so to speak, while in these cases the cause is externalized into the object that produces the effect. It is the very beginning of the process of constructing the mechanisms of physical causality, which will go on for the whole period of sensorimotor development.

These behavior patterns, that are so striking and frequent in children between five and eight months, were never seen in our gorilla. A wide variety of effect-producing objects (objects making sound, oscillating, rotating, producing flashes of light, etc.) were presented for many sessions to our subject: they required some simple action (like pushing or shaking) to produce their effect. The gorilla often did produce the action, did notice the result, but never tried to reproduce it by repeating more or less systematically her action. When she did repeat her action, she paid no attention to the result obtained. In other words, there was never any evidence that results "fed" action, which is the crucial property of secondary circular reactions, as it can be seen in children when, for example, the effort put in the action is varied at each repetition in order to obtain a variation in the effect. Objects presented were manipulated, but the schemata applied to them remained primary: they were taken, carried, moved, brought in front of the eyes, looked at, mouthed, and sometime thrown to the ground, i.e., they were always directly assimilated by body schemata.

However, both with these objects and with others, a more primitive sequential organization of schemata, which is also typically shown by children in this stage, could be detected. This is the successive application of different schemata to one and the same object, a sort of primitive but systematic exploration of object properties ("Sequences of primary schemata applied to objects" in Table 4).

The integration of schemata achieved in the preceding stage makes also possible the beginning of the construction of objects conceived as permanent entities existing in space independently from the subject's actions and perceptions: what Piaget calls "object-concept" or "object-permanence." This too is a long developmental construction which will reach true object-permanence only at the end of the sensorimotor period. However, indexes that objects are beginning to become permanent beyond the subject's actual perceptions are found in this stage. These indexes are: 1) the accommodation to rapid movement, 2) the reconstruction of an invisible whole from a visible fraction, 3) the search behaviors relative to objects that have been hidden in front of the subject.

Visual accommodation to rapid movements is the capacity to recover the position of an object when visual contact with it has been lost owing to its sudden displacement. An initial permanence is therefore maintained across brief interruptions. During the fifth month this capacity was still absent in our gorilla: when an object that she was visually fixating was dropped to the ground, she kept staring at the point where the object was (the hand of the experimenter that dropped the object). A few weeks later, she began to look to the ground after staring for some seconds at the point of disappearance. Finally at 168 days, as soon as the object was dropped, she immediately looked to the ground ("Visual accommodation to rapidly moving objects" in Table 4).

The second behavior typical of third-stage object-concept is that of recovering an object that has been hidden under a screen in full view of the subject, but so that a portion of it is still visible. With reference to its relevance to object-permanence, Piaget termed this behavior "the reconstruction of an invisible whole from a visible fraction": it indicated a higher degree of object-permanence in that the perception of only a part of the object is now sufficient to reconstruct the whole object. At five months, even if she was clearly interested in obtaining the object and tried to grasp it, she stopped any attempt to do so when the object was hidden so that only a portion of it (about a quart) could be seen. During the sixth month she would sometimes touch or slightly manipulate the screen, or go onto the visible portion of the object with her mouth, but never clearly grasp and recover it. Only at the very end of the sixth month she began to grasp the object with her hand and take it out of the screen ("Recovering partially hidden objects" in Table 4).

At the same time, however, if the object was completely covered by the screen, the gorilla did not make any attempt to search for it: she abruptly stopped any action and stared at the screen or around it, behaving as if the object had ceased to exist. Toward the middle of the ninth month, this behavior changed in a peculiar manner. When the object disappeared under the screen she went to look and search for it in the hand of the experimenter that was holding it ("Searching for hidden objects at the point of disappearance" in Table 4). Searching at the point where the

TABLE 5.
Stage 4 of Sensorimotor development in Gorilla

	Age in days Rm
MEANS-END COORDINATIONS OF SCHEMATA	
- Removing an obstacle to get to an object	258
- Leaving an object to grasp another	266
- Lifting a screen to recover a hidden object	356
- Pulling a string to obtain an attached object	368
OBJECT CONCEPT	
- Recovering a hidden object by uncoordinated attempts	310
- Recovering a hidden object by lifting a screen and grasping object	356
- Typical reaction	427

object was last seen is a typical behavior pattern of this stage, showing how much the existence of the object is still tied to the subject's perception and action: the subject goes to search where her last (visual) contact with the object was successful. Object-concept development in this stage appeared to be extremely similar to that of the child. On the other hand, the absence of secondary circular reactions represented a major divergence from the human pattern of cognitive development.

Stage 4

In stage 4 the sequential coordinations of schemata seen in the preceding stage develop into a more complex hierarchical organization of independent schemata. The principal manifestation of this organization is the coordination of schemata into a means-end relation. In this area, the development of the gorilla paralleled that of children. The establishment of some of these coordinations is reported in Table 5 ("Means-end coordinations of schemata"). It should be noticed, however, that only the most elementary of these coordinations, that of removing an obstacle (in this case, pushing away the hand of the experimenter) that is interposed between her and her visible objective (an object she wants to grasp), was manifested during the ninth month, as well as that of letting go an object that she held in order to grasp another. More robust coordinations began only at the end of the first year of life, that is, quite later than in children. In the domain of object-concept, the notion of a permanent object, existing in space independently from the subject's actions and perceptions makes substantial progresses during this stage.

The main index of this progress is the capacity to conceive objects as occupying a position in space even when they are not directly perceivable.

Such an ability is indicated by the recovering of an object that has been completely hidden under a screen in front of the subject: the subject reaches for the hidden object because he now knows that it has not vanished. Development in this domain also paralleled sequentially that of children, but here too it seemed to lag somewhat behind temporally.

Modifications of the behaviors showed in the preceding stage when the object disappeared completely under a screen occurred along the usually observed sequence. First the reaction of searching into the hand of the experimenter disappeared and, at the same time, the subject started to touch or manipulate the screen, without, however, succeeding in recovering the object. This occurred at 280 days.

Then the action of manipulating the screen became more prolonged and insistent (as if the subject knew that the object was somehow where the screen was), but the actions were uncoordinated and undirected to the goal of uncovering the object, so that success ensued only occasionally and more or less accidentally. First successful attempts of this kind were seen during the 11th month ("Recovering a hidden object by uncoordinated attempts" in Table 5). Finally, the object was recovered by a sequence of two clearly distinct actions: lifting the screen and grasping the object. This final coordination, that is considered the crucial achievement of stage 4 object-concept only occurred at 356 days, again quite later than what is usually seen in children. In fact, what seemed to be happening was a slowing down of the rate of development toward the end of stage 3 and in the passage to stage 4. The typical behaviors of stage 3 object-concept occurred during the sixth and seventh month, but, as it can be seen from Table 5, full stage 4 behavior patterns, both in means-end coordinations and in object-concept, occurred only toward the end of the first year of life, i.e. four to five months later.

This is further confirmed by the fact that the "typical reaction" of stage 4 object-concept was still present at the last date of systematic longitudinal testing, at 427 days. This means that our subject was still into stage 4 at 15 months of age.

This reaction shows the upper limit reached by object permanence in this stage. If, after having been hidden and successfully recovered under a screen A, the object is immediately hidden again under a different screen B, still in full view of the subject, the subject goes to search under the first screen. This happens because the object is not yet fully objectivized as existing independently in space. Its existence is, on the contrary, still dependent on the subject's action: thus, he searches where his immediately preceding action was successful, under the first screen. The subject's actions have still precedence over the objective spatial relations. Complete disappearance of this reaction marks, in fact, the achievement of stage 5 object-concept.

A large difference with respect to child development in this stage is the complete absence of coordinations of secondary schemata, that is those coordinations that construct the causal and dynamical relations of objects among themselves. In view of the absence of secondary circular reactions seen in the preceding stage, this result simply confirms the same tendency. Behaviors like dropping an object onto another, banging two objects together, etc. were never seen in our subject. Dynamic and causal relations among objects were in no way explored in the intense way which is typical of the children toward the end of their first year.

Further developments

In so far as they are a direct development of stage 4 secondary schemata coordinations, none of the "experimental" behaviors, like "tertiary circular reactions" and "instrumental" behaviors, which are the central core of stage 5, was seen in our subject, throughout the end of the testing sessions, at 15 months. For example, the gorilla showed no progress in solving the problem of recovering a stick from outside her cage by rotating the stick, so that its orientation would be parallel to the cage-bars, before pulling it in. In this case, her behaviors were similar to those of children when first presented with such a problem (see Piaget, 1974, pp.305–311, Obs.162–166), namely, pulling directly the horizontal stick against the bars and trying to force it through, but showed none of the gradual and systematic progresses that end up in anticipating the rotation of the stick, after grasping and before pulling.

Some very primitive form of secondary schemata was seen in this period. At 377 days the gorilla started banging a rattle against the floor, the cage walls and bars. Not much attention was paid to the effects, in terms of varying or graduating it; the action was, however, repeated and reinforced by the noise effect produced, and also applied to different substrates in sequence. At 427 days she began to slide a plastic ring against the floor by pushing it across the cage with her muzzle. She repeated this action several times and then applied it to other objects, including a wooden cup and a rattle, all making different noises when slid.

These, however, were the only two behaviors of this kind seen. They did not multiply and, especially, did not evolve into the more systematic effect-exploring patterns typical of "tertiary circular reactions", where the subject's action on objects is repeated by gradually and semisystematically varying its parameters to determine the correspondent variations in the effect.

DISCUSSION

The pace of initial sensorimotor development in gorilla closely mirrors that of the child. Stage 1 spans the first two months of life and stage 2 is completed during the sixth month. These temporal steps are almost identical to those of the child. These rates, however, progressively slow down in stage 3 and 4, so that cognitive growth lags more and more behind that of the child: stage 4 is achieved at the end of the first year of life and it is still in progress well into the 15th month, while children pass into stage 4 during the tenth month.

First stage development is identical to that of the child, both in the abilities developed and in their timing. So are the adaptations of sucking, vision and prehension developed during stage 2.

Schemata intercoordinations, on the other hand, though undergoing a sequence that parallels that of the child in both their order and their time of acquisition, show some interesting differences. After the development of hand-mouth coordinations (thumb-sucking and bringing grasped objects to the mouth), and before the development of any hand-vision coordination, the gorilla develops a buccal-prehension pattern, unlike anything seen in children. The ground for this development is the gorilla's development of independent locomotion during this stage (compare Table 2 to Table 3).

The capacity to reach objects by moving toward them appears toward the middle of stage 2, but, critically, is already present before the gorilla is able to grasp objects at sight. For this reason buccal prehension becomes the dominant pattern of interaction with objects in this period, completely vicariating hand-prehension.

The early establishment of this coordination has lasting consequences. It allows the development of a sort of "locomotive" pattern of object-exploration, consisting in moving toward the objects, moving around them, mouthing them, which is quite different from the one resulting from hand-prehension, where the object is grasped, taken to the eyes, turned around, and/or taken to the mouth: the subject moves about the object, rather than manipulating the object.

This fundamentally different pattern of interaction with objects does decline in frequency as soon as hand-prehension begins to be successful, and is finally displaced by it. It does not disappear entirely, as we have observed, but remains as a latent, more "archaic" pattern, alternative and in competition with the dominant one based on hand-prehension.

Stage 3 shows, on the other hand, a pattern of development widely divergent from that of children. Secondary circular reactions, the developmental cornerstone of this stage, were never shown by our gorilla. She did develop sequential coordinations of schemata centered on one object, but the schemata involved were always primary: no coordinations occurred between actions on objects and effects produced by the objects as a result of these actions. On the other hand, object-concept developed through the same sequence and at about the same ages as in children.

This divergence is noteworthy. It is not the lack of a general capacity of this stage: schemata coordinations do develop in the same fashion as in children, and they give rise to the sequential "explorations" of object properties and to the construction of object permanence. It is a specific type of coordinations which is absent: those where the object begins to play the role of an intermediary in a cause-and-effect chain. In secondary schemata the subjects acts on an object and the object produces an effect, and it is this relationship that is investigated in the repetition and variation of this cycle, i.e. through the "circularity" of secondary reactions.

Chevalier-Skolnikoff (1977) reports that secondary circular reactions did occur in the development of a gorilla infant she followed observationally. However, no description or example are provided and, furthermore, there are reasons to believe that this author has seriously misunderstood the concept of secondary circular reaction. Since in the same article she reports also observations of sensorimotor development in macaques, for which examples are provided, her study will be discussed and criticized in the following chapter, dealing with macaques' sensorimotor development.

The more complex coordinations of secondary schemata typical of stage 4, where object-to-object cause-and-effect relations begin to be constructed, were also not seen in our gorilla. On the other hand, hierarchical coordinations of independent schemata into means-end relation developed in this stage as in children, and so did object-concept.

Development in this stage was, however, slower than in children: object permanence was achieved at 356 days, i.e., at the end of the first year of life.

The difference with respect to children's development seems to involve the specific domain of physical cognition. Schemata did not explore causal and dynamical interactions among objects, but they were combined into a more advanced logical organization. Up to 15 months of age, none of the behavior patterns typical of stage 5 was seen. Some of the capacities that children achieve in stages 5 and 6 in specific domains of cognition (object-concept; physical causality, etc.) were, however, formally tested during the gorilla's second and third year of life: the results will be reported in the following chapters.

4 Early Sensorimotor Development in Macaques (*Macaca fuscata, Macaca fascicularis*)

Patrizia Potí
Istituto di Psicologia del CNR
Rome, Italy

INTRODUCTION

Macaques are among the most widespread nonhuman primates. They are diffused all over South and East Asia, from Afghanistan to China and from Japan to Timor, with one species present also in North Africa (Morocco and Algeria) and extending even to Gibraltar. Because of their excellent adaptability, they can live in a wide variety of habitats, ranging from tropical rain forest to grassland and dry areas, to the temperate woodland of Japan. For the same reason and since they are easy to keep and breed in captivity, they are the nonhuman primates most extensively used in laboratories for medical and psychological research.

The genus *Macaca* belongs to the family of Cercopithecidae, which together with the families of Hominoids, belongs to the Catarrhiny, or Old World monkeys (Schwartz, 1986). On the basis of biochemical (Sarich & Cronin, 1980; Baba et al., 1980), anatomical (Luckett, 1980) and fossil evidence (Napier, 1970; Hoffstetter, 1980; Pilbeam, 1984), it is assumed that the split within the Catarrhiny lineage between Cercopithecoids, to which all present day monkeys of the old world belong, and Hominoids, to which all extant apes and man belong, took place between 25 and 30 million years ago. Today, the only extant family of Cercopithecoids is that of Cercopithecidae, whose radiation, giving raise to present day genus, seems to have began about 15 million years ago.

The genus Macaca divides into 17 species. We will now examine in some detail the characteristics of the two species that are the object of this study: the Japanese macaque and the crab-eating macaque.

Japanese macaques (*Macaca fuscata*) live in the temperate forest of the Japanese islands, reaching 41N that is the northernmost latitude to which any nonhuman primate is found (Kavanagh, 1983). Their diet is based on fruits, leaves and crops.

Japanese macaques live in large groups of up to a few hundred members. The core of the group is constituted by dominant males and adult females with their babies, subadult males living at the periphery of the group.

Japanese macaques are quadrupedal and terrestrial with limbs of almost equal length. The tail is very short. Adult head-and-body length range is 535–607 mm in males and 472–601 mm in females. Adult body weight range is 11000–18000 gr in males and 8300–16300 gr in females. Body weight range at birth is 450–500 gr. The gestation period is 170–180 days and lactation lasts 6 months. The infantile phase lasts from birth to one year. The juvenile phase lasts from one year to the achievement of sexual maturity. Sexual maturity is reached at 4.5 years by males and at 3.5 years by females. Full growth is reached at about 10 years by males and at about 6 years by females (Napier & Napier, 1967). Life span is about 20 years.

Completion of brain growth is reached at about 3.5 years (Passingham, 1975).

Crab-eating macaques (*Macaca fascicularis*) live in the peninsula of Indo-China, the archipelago of Indonesia and the Philippine Islands. Their typical habitat is constituted by tidal creeks and mangrove swamps (Kavanagh, 1983). Their diet consists of fruits, leaves, crops, molluscs and crustaceans.

Crab-eating macaques live in groups ranging from 10 to 200 members. Adult dominant males and adult females with their babies occupy the core of the group with subadult males living at the periphery of the group.

This species is arboreal. Tail length is equal to head and body length. Adult head-and-body length range is 412–648 mm in males and 385–503 mm in females. Adult body-weight range is 3500–8286 gr in males and 2500–5680 gr in females. Body-weight range at birth is 230–470 gr. The gestation period is 167 days and lactation lasts 14–18 months. The infantile phase lasts about one year. The juvenile phase lasts until sexual maturity, which is reached at about 4.5 years in males and at about 3.5 years in females. Full growth is reached at about 4.5 years in females and 6–10 years in males. Life span is about 15 years (Napier & Napier, 1967). Completion of brain growth is reached at about 3.5 years (Passingham, 1975).

The present study investigated the cognitive development of Japanese and crab-eating macaques and aimed at providing basic data on the sequence of development of sensorimotor intelligence, as characterized by Piaget (1971a, 1974) in the human child. To this purpose, infant macaques were followed longitudinally since birth across the first four stages of sensorimotor development and compared stage by stage to children.

Although the cognitive abilities of macaques of various species have been extensively investigated in the past (see Jolly, 1972, for a review), very few studies are concerned with cognitive developmental data and the comparison to the human primate. Three of them, however, are directly relevant to our purpose. The first is an experimental study carried out by Wise, Wise and Zimmerman (1974). The authors investigated the development of object-concept in rhesus monkeys. Four subjects were longitudinally examined from birth to the fourth month. Behavioral indexes used were the same Piaget used to assess the achievement of the first five stages of object-concept in children. It was found that macaques passed through the same developmental sequence as children, but in a much shorter time. An observational study by Parker (1977) traced the behavioral development of an infant stumptail macaque (*Macaca arctoides*) living with its mother and a playmate from birth to six months. Through the observation of its spontaneous behavior data were derived concerning base structural organization of sensorimotor action, object-concept and imitation. With regard to object-concept results were quite similar to those obtained by Wise et al. (1974). A striking difference with respect to the human infant was instead found in the development of schemata coordinations: the macaque never displayed "secondary circular reactions", i.e., the initial development of the concept of physical causality, and, furthermore, it was extremely limited in combinatorial play. Chevalier-Skolnikoff (1977) observed the spontaneous behavior of three stumptail macaques (*Macaca arctoides*) infants living in a group of ten individuals from birth to six months. In this study, however, both data analysis and their interpretation contain some misunderstandings of the cognitive categories underlying the observed behaviors. By and large, results of these studies are in agreement with those of the present study and will be discussed in detail below.

SUBJECTS AND METHODS

All subjects tested were followed longitudinally since birth, though with partially different methodologies, depending on the setting in which they were raised.

Two infant Japanese macaques (*Macaca fuscata*), a male (Zu) and a female (Oi), were born and raised by their mothers in two cages of the Rome Zoo. Accordingly, they were followed only observationally, by recording their spontaneous behaviors. The male, Zu, was followed from birth to four months of age, and the female, Oi, from two weeks to four months. Only natural objects like pebbles, leaves, branches and mud were available in the cages and no external intervention from the experimenters was attempted. The behavior of each infant was recorded by paper and pencil in one-and-a-half-hour long sessions, with an average of two sessions per week.

A third Japanese macaque, an infant male (Pn), born in the Rome Zoo, was instead hand-reared since birth by a human caretaker. In this case, where the experimenters could freely interact with the subject, the usual piagetian methodology was followed: the subject was given a variety of objects to act upon, and it was both observed while freely playing with them and prompted to perform specific tasks. Data are drawn from one-to-two hour experimental sessions, that occurred twice a week for the first month of life and once a week afterwards. Sessions did not begin at birth, but only in the infant's second week of life and lasted until the end of the fourth month. They were all videotaped and then transcribed. A 20 minute videotape (3/4", black and white reel) summarizing the development of object-concept is available. The fourth subject was a crab-eating macaque (*Macacafa scicularis*), an infant male (Gu), that was hand-reared from birth by a human caretaker. The same methodology of the preceding subject was followed. Testing took place in hour-long sessions, that began on the day of birth and extended to the fourth month. A finer testing grain was here adopted: sessions occurred every day for the first two weeks of life, then three times a week, and twice a week toward the end of the study. All sessions were videotaped and transcribed. Original 3/4" color cassettes (U-MATIC, PAL system) are available.

RESULTS

Cognitive development of our subjects was analyzed in terms of the sequence of stages of sensorimotor intelligence described by Piaget (1971a, 1974) for the human infant. The analysis reports, for each subject, the time of appearance, duration and specific characters of each of the first four sensorimotor stages and compares them to those of children. The behavioral items identifying each of the four stages, kept as much as possible constant in the studies of the four species, were grouped into tables (Tables 1–5) that report, for each of them, and for each subject, the age in days at which it was first observed. Only for the last item of the last table (Table 5), "Typical reaction", the age is that of the last testing session in which the behavior was still observed.

It should be noted that because of the reported differences in the methodologies followed and in the timing of testing, not all the same behavioral indexes are available for all subjects. As we shall see, however, the general developmental picture appears to be quite stable across the four subjects.

TABLE 1.
Stage 1 of Sensorimotor Development in
Macaques

	Age (days)	
	Zu	Gu
Adaptations of Sucking		
-Orienting to contact within mouth-area		3
-Orienting to contact outside mouth-area		6
Adaptations of Vision		
-Partial visual pursuit	8	6

Stage 1

In the first stage of human cognitive development, from birth to the end of the first month of life, reflex activities present since birth develop, through assimilating external stimuli and progressively adapting to them, into sensorimotor schemata. The schemata of this stage are "primary": they directly assimilate "objects" through the main organs of the mouth, the eye and the hand. The main characteristic of this stage is the fact that these primary schemata remain separate from and uncoordinated to each other. Developmental progress consists in their increasing adaptation and differentiation.

Like the human infant, macaques showed at birth the reflexes of sucking and rooting, i.e., an undirected search for the nipple performed by lateral movements of the head and the mouth. Sucking was indiscriminately applied to anything that came in contact with the mouth. Both these reflexes became progressively adapted. As in children, the search for the nipple became oriented by contact, i.e., the head turned immediately in the direction where contact occurred, so that the mouth could center the nipple without attempts. First, this orientation occurred only when the nipple touched the mouth area; then it generalized and occurred also when the nipple touched the cheek or the chin. This development (which could be tested only in one of the hand-reared subjects) occurred during the first week of life (see Table 1).

Visual fixation of strong stimuli (a light or a colored object held in front of the eyes) was initially confined to a few instants: as soon as the stimulus began to move slowly, visual contact was lost. Gradually the infants became capable of visually following this slow moving stimulus. This adaptation occurred at the beginning of the second week: objects slowly moving along an arc in front of their eyes could be followed, but for no more than a 60 degree arc, after which contact was lost ("Partial visual pursuit" in Table 1).

Also prehension activities closely resembled those of children. Grasping reflex was very effective since birth, and it allowed the macaque infants to cling to their mothers' fur or to the maternal substitute. But when supported, macaques showed the typical random and repetitive movements of opening and closing fingers or waving arms in the air of this stage. As in children, grasping activated by contact with the palm of the hand was not accompanied by holding (except in clinging to the fur, as observed above, which is, however, activated by absence of body support): objects grasped are immediately released.

In this first stage macaque infants underwent the same developments seen in human infants, but at a much quicker rate. In fact, stage 2 abilities began to appear already during the second week of life, as compared to the second month of the child.

In one area, however, development is very different, that of locomotor abilities. As it can be seen in Table 2, already at birth the macaque can fully support its head erect, and it can roll from supine to prone position two days later. Most important, it begins to move independently in space at 5 days, by creeping on the floor, and, at the same time, it stands erect on the four limbs. It can walk confidently on all fours by the beginning of the second week and climb on a fence few days later. This represents not only an acceleration of the absolute rate of development with respect to the human infant, as seen for the whole of stage one, but also, and most crucially, a displacement in the relative time of onset of the different abilities. The human child, in fact, begins to be able to move independently in space at about 8 months, that is when he is in the fourth stage of his sensorimotor development, while the macaque does so at the beginning of the second stage.

Stage 2

In the second stage of human cognitive development primary schemata centered on the mouth, the hand and the eye continue to adapt and differentiate. The infant becomes capable of visually fixing and following moving objects without losing contact. Sucking differentiates into a number of non-functional schemata, unrelated to the feeding activity, like the rhythmical protrusion of the tongue or finger-sucking. Grasping develops into a number of schemata for tactile and kinesthetic exploration: touching and fingering, scratching and holding.

But the most important achievement of this stage is the development of intercoordination among the separate schemata centered on the different organs: it is only through these intercoordinations that the separate oral, tactual, and visual worlds of the preceding stage become unified into a single one and hence begin to construct "objective reality".

Adaptations of sucking and grasping occurred during the second week,

TABLE 2.
Development of Locomotion in Macaques

Phase	Behavior	Age (days)		
		Zu	Pn	Gu
1	Supporting head erect	1		1
	Clinging to fur	1		1
	Rolling to prone			2
2	Standing on forelimbs	3		4
	Creeping			4
	Climbing on fur			7
	Walking with flexed hips	5		5
3	Walking on all fours	8	7	7
	Climbing on fence	10	10	9

as reported in Table 3, and are not different from those seen in children.

Adaptations of vision and coordination among sucking, prehension and vision showed instead a different developmental progression. As in children, the first coordination that became established was that between the hand and the mouth, manifested in thumb-sucking. As we saw above, however, at the beginning of the second week the macaques were capable of moving in space. This capacity gave rise to a coordination that is never seen in children: beginning from the seventh day of life, on seeing an object in front of them the macaques moved toward it and grasped it with their mouth. In other words, macaques developed the same pattern of "buccal prehension" described, in the preceding chapter, for the gorilla. For the

TABLE 3.
Stage 2 of Sensorimotor Development in Macaques

	Age (days)			
	Oı	Zu	Pn	Gu
Adaptations of sucking				
- Tongue protrusion		8		
Adaptations of grasping				
- Holding an object for more than an instant				7
- Tactual investigation of one's own body		8		
- Scratching on surfaces		12		8
Adaptations of vision				
- Complete visual pursuit		12		10
Coordinations				
- Thumb sucking			7	6
- "Buccal prehension"		8	8	7
- First unsuccessful attempts to visually guided hand prehension			9	9
- First successful attempts to visually guided hand prehension			13	14
- Fully developed visually guided hand prehension	24	28	22	26

same reason, and at about the same time, visual pursuit of moving objects developed into a "visual-locomotor" coordination, also never seen in children: the macaque would visually track moving objects by walking after them.

In the following period, buccal prehension became the dominant mode of interaction with objects. Macaques did not develop a prehension-sucking coordination, i.e., bringing a grasped object to the mouth or grasping with the hand an object hold in mouth, as children do, after hand-mouth coordination. Instead, they went on to develop vision-prehension coordination, i.e., the capacity to grasp objects with the hand at sight.

The establishment of vision-prehension coordination followed the usual course seen in children. It began with the first (unsuccessful) attempts to grasp at sight, followed by successful attempts when the hand and the object are perceived within the same visual field, that is, by visually guiding and adjusting the movements of the hand toward the object. It then went on to the final achievement of prehension-vision coordination, when objects are grasped with a single, precise movement of the hand from outside the visual field, that is not visually guided and adjusted (see Table 2).

Buccal prehension, however, remained dominant for this whole period. In fact, failures to grasp at sight would be usually followed by grasping with mouth. A peculiar pattern, that we labeled "substitute prehension", was also seen during this period: on failing to grasp an object at sight, the macaque would carry the empty hand to the mouth and suck it while keeping staring at the object. These "substitute prehensions" decreased with increasing success in grasping objects at sight and disappeared after prehension-vision coordination was completed, thus confirming their "substitute" nature.

What did not disappear, even after the infants were perfectly capable of grasping at sight in any condition, was buccal prehension. It remained an important pattern of accessing objects for the whole period of observation, side by side and equally concurrent with that of hand prehension.

Stage 3

The attainment of fully developed vision-prehension coordination marks the transition to the third sensorimotor stage in human development. From now on schemata begin to be coordinated in sequences leading, in stage 4, to what Piaget calls "intentional adaptations", that is the hierarchical subordinations of means to an end. Precursors to this hierachical organization are the simpler sequential coordinations of schemata found in this stage. They are of two kinds: sequential application of different schemata to the same object and "secondary circular reactions". All subjects manifested the first type, beginning from the 6th week (see Table 4). For

TABLE 4.
Stage 3 of Sensorimotor Development in Macaques

	Age (days)			
	Oi	Zu	Pn	Gu
- Sequences of primary schemata applied to objects	38	44	38	39
- Secondary circular reactions	-	-	-	-
Object Concept				
- Visual accommodation to rapidly moving objects	38	41	28	33
- Recovering partially hidden objects			38	39
- Searching for a hidden object at the point of disappearance			42	45

example, they grasped an object at sight and brought it to the mouth, then took it away from the mouth, held it and observed it, then brought the object back into the mouth again. Such sequences were furthermore reiterated many times.

On the other hand, macaques, just as the gorilla, never showed any secondary circular reaction. Both subjects followed observationally were once seen performing one of the typical actions children use to obtain secondary effects, namely shaking an object. In the 6th week they briefly shook a branch they were holding. However, they did not pay any attention to the effects of shaking the branch, namely its oscillatory movements, and hence did not try to replicate or vary such effects by repeating their action on the branch.

In the interactive studies, the experimenters tried, from the 4th week on and always without success, to elicit secondary circular reactions employing a variety of objects producing different effects when touched or otherwise manipulated. Objects and devices that would move, swing, rotate, produce sound or noise or light when activated by simple actions (like pressing). Subjects, however, acted on these objects in their usual way, that is, by mouthing, grasping, touching or climbing on them; even when their actions did produce the intended effect, they were never much interested in it and did not repeat the action in order to reproduce the effect. The same thing happened when the experimenters demonstrated the effects.

The absence of secondary circular reactions represents a major difference in the developmental sequence of macaques as compared to the human child.

On the other hand, the development of the concept of object (or "object-permanence"), which begins in this stage, was quite similar to that of children. The same behavioral indexes that characterize human object-concept development in this stage could be detected in macaques (see Table 4).

Between the 5th and the 6th week, macaques displayed "visual accommodation to rapid movements" i.e., the capacity of recovering visual contact with a moving object, when this contact has been lost because of a sudden displacement of the object. Before this age, when an object held in front of the subject was dropped to the ground, the subject continued to stare at the point where the object was. From this point on, instead, the subject was able to recover the unseen trajectory and look to the ground where the object had fallen.

Two further behavioral indexes of stage 3 object-concept development could be ascertained only with the two hand-reared subjects. They refer to the behaviors shown when an object is covered with or disappears under a screen while the subject is looking at it.

In their 7th week, subjects became capable of "reconstructing an invisible whole from a visible fraction": they could recover the object, if it had been only partially covered, so that a portion of it remained visible (Table 4, "Recovering partially hidden objects"). At the same time, however, if the object was completely covered, subjects stopped their actions, looked around, but did not make any attempts to search for it under the screen (as if the object had vanished).

In the 8th week, also the behaviors with respect to totally hidden objects changed in the peculiar manner that is typical of this stage. Rather than losing interest when the object disappeared, they went to search in the hand of the experimenter which had moved the object to under the screen. They searched for the object at the point where they last saw it before its disappearance (Table 4, "Searching for a hidden object at the point of disappearance").

Stage 4

It is in this stage that sequences of independent action schemata hierachically organized into means-end relations make their appearance. Macaques began to show clear instances of such coordinations in their 10th week. The time of appearance of some of these means-end coordinations of schemata, such as lifting a screen in order to grasp an object or pulling a rope in order to grasp an attached object, are reported in Table 5. Many such coordinations, however, were observed throughout the stage. In the domain of object-concept, the notion of a permanent object, existing in space independently from the subject's actions and perceptions, developed as seen in children of this stage.

Such progresses began, from the 9th week, with the disappearance of the behavior of searching in the experimenter's hand when the object disappeared under the screen and the parallel appearance of attempts to recover the object by uncoordinated and undirected manipulations of the

TABLE 5.
Stage 4 of Sensorimotor Development in Macaques

	Age (days)		
	Oi	Pn	Gu
Means-End Coordination of Schemata			
- Leaving an object to grasp another	79		
- Lifting a screen to recover a hidden object		63	67
- Pulling a string to obtain an attached object		70	73
Object-Concept			
- Recovering an hidden object by uncoordinated attempts		56	59
- Recovering a hidden object by lifting the screen and grasping the object		63	67
- Typical reaction		98	87

screen. The subjects touched and moved the screen (a soft piece of cloth), and even jumped or scrambled on it. The accidental displacements of the screen thus produced sometimes succeeded in uncovering the object that got recovered (Table 5, "Recovering a hidden object through uncoordinated attempts").

Then, and, significantly, when real means-end coordinations appeared, the object began to be recovered by a precise sequence of two actions: lifting the screen and grasping the object. Instances of the "typical reaction" of this stage, when the object is hidden consecutively under two different screens, were seen in both subjects till the end of the testing sessions and attested that in the 13th week (Gu) and in the 15th week (Pn), they were still into stage 4.

The developments just seen are essentially the same seen in children's stage 4. What was not observed in macaques were coordinations of secondary schemata that are also typical developments of stage 4 in the human child. Given the absence of secondary circular reactions in stage 3, this result is not surprising. Coordinations of secondary schemata are the essential way of constructing cause and effect relations among objects, i.e., the notion of an objective physical causality as distinct from subjective efficacy. They are also essential in exploring and determining most of the physical and dynamical properties of objects. Schemata like sliding an object against a surface or banging an object against another were never seen in macaques. Objects were explored by sequences of schemata as in the preceding stage, and during this stage new schemata made their appearance, such as sliding the palm on an object, sometimes while holding it with the other hand, and digging earth with hands, but all these schemata remained primary, i.e., centered on the direct assimilation of the object to some body part of the subject, and hence useless from the point of view of constructing physical causality.

Further developments

The increasing differences of macaques' development with respect to children makes it difficult to talk, in any meaningful way, of an eventual stage 5 or 6 of sensorimotor development. The central features of stage 5, "tertiary circular reactions" and the active experimentation of object-to-object relations, were never seen in any subject till the end of the testing sessions, and, in one case, even for a long time afterwards.

In fact, one of the subjects, Oi, was occasionally observed and tested for one and a half years after the termination of the systematic study. It was, furthermore, one of the subjects of the experimental studies on the stick and the support problems and on object-concept, directly relevant to the abilities of stages 5 and 6, which will be reported in the following chapters. The only more complex schema relating objects during this period was first seen at 10 months: the macaque started rubbing objects against a substrate (the floor or the walls of its cage), i.e., a behavior that would have been typical of stage 4 secondary schemata coordination, as explained above. But this remained a single instance, and was not accompanied or followed by other such coordinations.

DISCUSSION

The first thing to notice in comparing sensorimotor intelligence development of macaques to that of children is the large difference in the rates of such development. Children pass into stage 2 toward the end of their second month of life, macaques at the beginning of their second week. Stage 3 and 4 are reached by children during their sixth and tenth month, respectively; toward the end of their first month and the beginning of the third, by macaques. These data are perfectly consistent with those of Parker (1977), who followed longitudinally the sensorimotor development of an infant stumptail macaque (*Macaca arctoides*). Parker found the beginning of stage 2 in the second week, that of stage 3 in the fourth and that of stage 4 in the ninth.

Though a more formal comparison, across all species tested, will be drawn in chap. 6, one can also see that, even staying much faster, the macaques' developmental rates progressively decrease in successive stages of development with respect to children's, as if approaching a limit.

In the second place, the course of development shows a pattern of increasing differentiation from that of children. Differentiation and adaptation of initial reflex and their development into primary schemata during stages 1 and 2 are quite similar to those of children. On the other hand, the development of schemata intercoordinations in stage 2 already shows a different course. Macaques do not develop a prehension-sucking coordination before prehension-vision coordination as children do. Owing to their very early acquisition of locomotion, at the very beginning of stage 2, they develop a "visual-locomotive" coordination leading to a pattern of prehension-by-mouth which is never seen in children. When grasping objects at sight with the hand is still impossible, because of the lack of coordination between eyes and hand schemata, the macaque can grasp objects at sight with its mouth, by moving toward them. In fact, since eyes and mouth are fixed with respect to each other, this is an easier coordination to achieve than that between eyes and hand. The hand moves through independent schemata and, to be guided by vision, its coordinates in space need to be mapped into those of the visual field (a process that takes the whole of stage 2 to develop), while any direct approach to the object through the visual field will automatically target the mouth onto it.

Though this pattern is the same as that seen in the gorilla, because of their very early acquisition of locomotion capacity, macaques develop it at an earlier time, i.e. at the very beginning of stage 2. Its further evolution and fate is, therefore, different than in the gorilla. In fact, when hand prehension will be fully developed, and this happens after quite a time, in terms of the macaque's fast rate of development (about two weeks), this pattern will be well established and will not be replaced. It continues, even after the development of hand prehension, and will consequently restrict the type and number of situations in which hand prehension will be exercised. However, as discussed above in the gorilla's case, buccal prehension determines a pattern of interaction with objects that is fundamentally different from that determined by hand prehension. The persistent concurrence of these two prehension patterns, therefore, will have consequences on the type and frequency of object-exploration patterns subsequently employed that, in the macaque, will be much more extreme than in the gorilla.

The central behavior pattern of stage 3, secondary circular reactions, is entirely absent from the development of macaques. All efforts to elicit secondary circular reactions with a variety of effect producing objects failed. This absence was also noted in Parker's (1977) study, quoted above:

> The stumptail's behavior patterns during this stage . . . were not reciprocally coordinated with object sounds and motion (they did include object manipulations) and they were not reinforced/released by perceived contingency of the object's behavior on his preceding behavior . . . he never repeated his actions on a rattle, or a swinging toy, or other objects I gave him, except in biting on them or climbing on them. (p. 66)

This description applies equally well to all our subjects; the schemata through which objects were manipulated, though increasingly more numerous and differentiated, remained primary, i.e., the objects were directly assimilated by body part schemata. This distinction is extremely important, because secondary schemata, by coordinating action on an object with the effects produced by the object itself (and not directly by the subject's action), represents the first step in the construction of objective cause-and-effect-relationships, i.e. the fundamental notion of the physical domain of cognition.

Chevalier-Skolnikoff (1977), on the other hand, claims the presence of secondary circular reactions in three infant stumptail macaques observed by her. Her study is, however, fraught with misunderstandings. First of all, behaviors are analyzed according to sensory modality and body part, a classification that has no theoretical justification, runs contrary to the very notion of schema as an integrated behavior pattern and furthermore becomes completely meaningless after the establishment, in stage 2, of schemata intercoordinations. Thus, one does not know what to do with statements like "secondary circular reactions have been observed only in the tactile/kinesthetic, visual/body modalities" (p.172). If specific examples of secondary circular reactions are looked for, then the followings are found: "Repetitive interanimal mouthing and play-biting are the most prevalent secondary circular reactions" (p.172). It is hard to see how these behaviors could conform in anyway to the physical causality exploration patterns of typical secondary circular reactions. A third behavior quoted, branch-shaking, could qualify as such, if, for example, attention was being paid to the oscillatory movements of the tip of the branch and the repeated actions would show signs of being regulated by the movements themselves (like shaking more or less quickly, or intensively, as a function of the larger or smaller, quicker or slower movements of the tip of the branch). A subsequent comment of the author, however, makes clear that this is not the case: "Branch-shaking was accomplished kinesthetically [which, presumably, means without visual control of the effect] and was generally received as a visual/body communication by others" (p.172).

In the development of object-concept, instead the macaques' sequence was identical to that of children. Given the absence of secondary circular reactions and their related patterns, stage 3 temporal duration and boundaries were, therefore, defined in terms of the object-concept developmental sequence. This divergence in the developmental pattern of macaques with respect to children persists and gets amplified in stage 4. On the one side, object-concept continues to develop as in children (and the appearance of its typical behaviors was, therefore, used to define stage 4, as well), but, on the other side, coordinations of secondary schemata, which begin to construct physical and dynamical relations among objects themselves, and are manifested in the typical "exploratory" object play (throwing, banging, stacking, etc.) of the children of this and subsequent stage, did not develop at all. It should, however, be noted that complex sequential coordinations of schemata did develop. Specifically, the hierarchical organization of schemata into means-end relation, which is also characteristic of this stage, was present and widespread.

As in the gorilla's case, it seems, therefore, that what is lacking is not the capacity for complex coordinations in general, but only of coordinations relative to the physical domain of cognition.

Given the central role played by the development of these coordinations in the next, fifth, stage, through tertiary circular reactions and systematic instrumental behaviors, and given their absence throughout the end of testing, there was no meaningful way of characterizing an eventual stage 5 development in macaques. Rather, attempts to test the presence of abilities discriminating between the different domains of cognition, which appear together in children at stage 5 and stage 6, were made with older subjects and will be reported in the following chapters.

5 Early Sensorimotor Development in Cebus (*Cebus apella*)

Giovanna Spinozzi
Istituto di Psicologia, C.N.R.
Rome, Italy

INTRODUCTION

The capuchins are among the most numerous and widespread nonhuman primates of the New World, widely diffused in Central and South America from Belize to northern Argentina. They constitute the genus Cebus whose taxonomic position and phylogenetic origin, as in general that of New World primates, is still today a matter of debate.

Most recent consensus (see, Ford, 1986; Schwartz, 1986) has all the New World primates, or Plathyrrhini, divided into three families, Cebidae, Callithricidae and Atelidae, with Cebus belonging to the first one. Plathyrrhini share a common ancestor with Catharrhini, or Old World monkeys, in Africa, where they originated during the Eocene, approximately 45 million years ago. This common stock differentiated into a more conservative group, which retained three premolars, the New World ancestral monkeys, and a more progressive one, with two premolars, the Old World ancestral monkeys. The conservative group reached South America across the South Atlantic, probably by rafting and island hopping, when the distance between Africa and South America was smaller than today (Ciochon & Chiarelli, 1980).

It is noteworthy that, of all the extant Plathyrrhini lineages, Cebus was the first to separate from a common ancestor, at the beginning of the late Oligocene, and has thus undergone the longest separate evolutionary history. Hence, it is phylogenetically the most distinct genus of New World primates. Four species are recognized within the genus Cebus: *C. apella, C. albifrons, C. capucinus, C. nigrivittatus. Cebus apella,* called tufted or black-capped capuchin, which is the subject of the present study, is distinguished by the characteristics from which it derives its common names. It is an arboreal primate living in the canopy of the tropical forest. It is the most common and widespread of the four species, ranging from Colombia to the Atlantic coastal forest of Brazil, to northern Argentina. It is omnivorous, mainly feeding on fruits and insects. It lives in small social groups of 7–30 individuals, with more female than male adults.

Its foraging behavior displays unusual and complex manipulative abilities, in exploiting food sources not easily accessible. It breaks off twigs, shapes them with its teeth and hands and inserts them under the bark of trees in order to extract small insects (Thorington, 1967). It strikes hard-shelled fruits against hard surfaces in order to break them open and gain access to their content (Freese, 1977; Izawa, 1979; Izawa & Mizuno, 1977); occasionally, for the same purpose, it even uses a detached object, like a stone or a piece of wood, as a pounding tool (Struhsaker & Leland, 1977; Antinucci & Visalberghi, 1986). Among nonhuman primates, such behaviors are only found in chimpanzees (Struhsaker & Hunkeler, 1971; Sugiyama, 1981; Boesch & Boesch, 1983).

Body weight in adults ranges from 3 to 5 kg, with males outweighting females by 1 or 2 kg. Both males and females reach sexual maturity at about 4 years of age, but become fully adult at about 8. In females, the length of gestation is about 180 days and the menstrual cycle varies between 15 and 20 days (Napier & Napier, 1967). No reliable physical maturation data are available on this species.

The infants are strongly dependent on their mother until 5 months of age and are almost constantly carried by her. Then, this physical dependence becomes progressively weaker even if at 1 year the infant is still given milk while resting on its mother. After this age, which marks the beginning of the juvenile phase, the little cebus is rarely carried on its mother's back and starts engaging in social play with other members of its age class (Izawa, 1980).

The comparatively advanced cognitive abilities of cebus were discovered by Kluver (1933) in his classic studies devoted to investigate the cognitive mechanisms underlying various forms of intelligent behavior in some species of New and Old World monkeys. Kluver found that cebus (*C. apella*) was superior to other monkeys (*Macaca fascicularis*) in various types of problem solving tasks and moreover some individuals were particular proficient in using and manipulating a wide variety of objects as tools in different experimental situations. From that time, practically no attention has been paid to the cognitive capacities of this primate species.

Recently, Mathieau et al. (1976) tested object permanence in this species (see, however, chapters 7 and 8 below, for a critique). Westergaard and Fragaszy (1987), Antinucci and Visalberghi (1986), Visalberghi (1986) investigated several types of tool use abilities. Parker and Gibson (1977) reviewed most of the (prevalently, unsystematic) observations of complex cebus manipulations and tool uses and argued that in this field cebus displays abilities similar to those of the great apes. The aim of the present study is to provide baseline longitudinal data on the cognitive development of cebus. Three *Cebus apella* were followed since birth and tested for their sensorimotor intelligence development.

SUBJECTS AND METHOD

Subjects were three *Cebus apella*, two females and one male, born in a colony housed in our lab. The three infants Carlotta (Ca, born February 1984), Roberta (Ro, born July 1986) and Pepe (Pe, born March 1987) were separated from their mother at birth and hand-reared in a house by a human care-taker.

As they grew up and were completely weaned, they were reintroduced into a group of conspecifics housed in our lab which included one adult male, one subadult and one juvenile female: Carlotta was reintroduced at 8 months of age, Roberta at 6 months and Pepe at 5 months. Because of their acquired familiarity with humans, the subjects could, however, be tested for further development even after their reintroduction, by temporarily removing each of them from the group with no stress.

The infant cebus were observed and tested in hour-long sessions. Sessions began on the day of birth and occurred twice a week for the first month and once a week afterwards. In each session, subjects were given objects and toys specifically chosen to elicit behavioral responses characteristic of the different sensorimotor stages. These included: sticks, rubber puppets and balls, plastic keys and rings, chains, strings, plastic rattles, nursing teat. The usual "clinical" methodology of piagetian studies was followed, freely combining observation of spontaneous interactions and provoked responses.

All sessions were videotaped and then transcribed and analyzed for each behavioral episode. Original 3/4" color cassettes (U-MATIC, PAL system) are available.

RESULTS

Data analysis is based on the identification in cebus of those behaviors described by Piaget (1971a, 1974) as characterizing the first four stages of sensorimotor development in the human infant.

Tables 1, 2, 4 and 5 show, for each of the four stages, respectively, the age in days at which each subject showed for the first time the behavioral items that are a crucial manifestation of the stage abilities. For the last item of the last table (Table 5), "Typical reaction", the age reported is that of the last testing session where the behavior was still observed. Furthermore, behavioral parameters relative to the acquisition and development of independent locomotion are reported in Table 3, because, as we already saw in the preceding studies, they seem to interact with the course of cognitive development.

Stage 1

Like the human infant in this stage, which spans its first month of life, cebus displayed at birth simple reflex activities relative to vision, suction and prehension that evolved into primary schemata through repetition and progressive adaptation to external stimuli. Most of these adaptations constitute a continuous process leading into stage 2 without any clearly markable behavioral landmark. A few of the most characteristic ones are described in what follows.

With regard to vision, in the first week of their life all three subjects were unable to fixate and follow with their glance a slowly moving strong visual stimulus, such as a brightly colored object or the flame of a match, located in front of their eyes. Like the human infant, they could only fix for a few instants the stimulus but lost contact as soon as it moved.

During the second week, they began to follow the same stimulus slowly moving along a circle in front of their eyes, but for no more than a short arc (about 60 degrees), after which they lost contact and were unable to reestablish it ("Partial visual pursuit", in Table 1).

As in the human infant, suction began with rooting and sucking reflexes that were present since birth: the mouth tapped repeatedly with a continuous and circular movement any surface with which it came into contact and tried to suck. First discernible adaptations occurred during the second and third week. Sucking became correctly oriented by contact: if the nipple contacted the mouth area, upper or lower lip, right or left, the head immediately turned in the right direction and centered the mouth on the nipple ("Orienting to contact within mouth-area", in Table 1).

Later on this adaptation extended to a wider area: the head turned in the right direction even when the nipple contacted either cheek, nose or chin ("Orienting to contact outside mouth-area", in Table 1).

The grasping reflex was also present since birth: any tactile stimulation of the palm caused the hand to close. Grasped objects, however, were not held: the hand grasped and soon after let go, if contact persisted then grasping was renewed and so on. Arm movements were impulsive and undirected.

TABLE 1.
Stage 1 of Sensorimotor development in Cebus

Behaviors	Age in days		
	Ca	Ro	Pe
VISION			
- Partial visual pursuit	10	10	11
SUCKING			
- Orienting to contact within mouth-area	10	14	6
- Orienting to contact outside mouth-area	15	23	11

All these developing schemata remained, throughout this stage, un-coordinated between each other: sucking, for example, became oriented by contact with the nipple, but if the nipple was only seen, no response was activated. Likewise for grasping. The three worlds that are being organized by the eyes, the mouth and the hand remain separate and un-connected.

During all this period, cebus were also completely unable to locomote or move in space. They could support their head erect and could turn themselves from supine to prone position (Table 3).

Stage 2

Two lines of development characterize stage 2: the increasing adaptation and differentiation of the primary schemata relative to vision, prehension and suction and, most important, their progressive intercoordination cul-minating, at the end of the stage, into the construction of a unified sen-sorimotor universe. At approximately the fourth week, the infants became capable of following a slowly moving visual stimulus throughout a 180 degrees arc, both forward and backward, without losing visual contact ("Complete visual pursuit", Table 2).

Prehension became sustained: a grasped object was no more immedi-ately released but held in the hand. Finally, out of the sucking schema there developed the systematic protrusion of the tongue, as a non-func-tional, but recognitory schema: when the bottle nipple was visually per-ceived but not touched by the mouth.

Coordination of prehension, suction and vision followed a somewhat different course than that of the human infant, but especially than that of the other nonhuman primate species tested. In the standard human se-quence prehension-suction coordination precedes prehension-vision co-ordination. First hand-mouth coordination is established, as manifested in bringing the hand (usually, the thumb) into the mouth, then grasped objects are taken to the mouth, or objects hold in mouth are taken away by the

TABLE 2.
Stage 2 of Sensorimotor development in Cebus.

| | Age in days | | |
	Ca	Ro	Pe
ADAPTATIONS OF VISION			
- Complete visual pursuit	19	32	26
ADAPTATIONS OF SUCKING			
- Tongue protrusion	19	27	26
ADAPTATIONS OF PREHENSION			
- Holding an object	19	27	26
COORDINATIONS			
- Thumb sucking	27	37	32
- Bring to mouth grasped objects	37	42	46
- First unsuccessful attempts to visually guided			
hand prehension	27	37	32
- First successful attempts to visually guided			
hand prehension	37	42	46
- Fully developed visually guided hand prehension	52	63	68

hand; only at this point the infant begins to attempt (unsuccessfully) to grasp objects that he sees in front of him.

In cebus thumb-sucking appeared toward the end of the first month. Then, and simultaneously, both attempts to bring to mouth grasped objects and to grasp at sight began. Neither of them was successful for a while, but both developed in a parallel fashion. It was only when grasping at sight began to become successful, which happened, as in the human infant, through visually adjusting the hand when it enters into the visual field, that grasped objects were successfully taken to and from the mouth (see Table 2). Finally, the last phase of prehension-vision coordination, as manifested by the ability to grasp objects at sight without visually adjusting the hand when it enters into the visual field, was achieved.

In view of these differences and for comparative purposes, it is important to look closely at the development of locomotion during this stage, which also differed from that of the other two species tested. As it can be seen in Table 3, the infant cebus were not capable of moving autonomously in space for the whole period during which prehension coordinations were unsuccessful: it was only after prehension-suction coordination was established and prehension guided by vision began to succeed that subjects could perform locomotor movements, even if the hind legs were still dragged on the ground. Consequently, the peculiar pattern of "buccal prehension" found in the other two species during this period did not appear in cebus. Some instances of buccal grasping guided by sight were found, but only in the case of the bottle nipple when it came close enough to the mouth to be reached by head movements. Full locomotive abilities appeared at the

TABLE 3.
Development of Locomotion in Cebus.

Behaviors		Age in days		
		Ca	Ro	Pe
Phase 1	- Supporting head erect	4	4	4
	- Clinging to fur	4	4	4
	- Rolling to prone	4	4	4
Phase 2	- Creeping	27	37	26
	- Climbing on fur	27	37	32
	- Walking with flexed hips	38	49	46
Phase 3	- Walking on all fours	52	70	74

same time when prehension-vision coordination was completely achieved. Even during this period, however, no systematic pattern of buccal prehension ever developed.

Stage 3

The achievement of primary schemata intercoordinations marks the transition to the third sensorimotor stage. Two developmental progresses can be followed in this stage. On the one hand, the beginning construction of an "object-concept", i.e. of the notion of an object that exists in space as such, independently of the subject's actions and perceptions. On the other hand, the beginning construction of objective causality, i.e., of cause and effect relations holding among objects themselves. Of course, the word "beginning" needs here to be stressed: both these developments will go on for the whole sequence of the six sensorimotor stages.

Development of object-concept follows a course remarkably similar to that of the human infant. As in children, the first behavioral index of this development was the appearance of "visual accommodation to rapid movements", i.e., the capacity of tracking the trajectory of a moving object when visual contact with it has been lost because of a sudden displacement. A suspended object was dropped to the ground in front of the subject and while up to this moment the subject continued to stare for a while at the point of disappearance, it now quickly turned its glance to the ground, thus showing that it could reconstruct the unperceived trajectory of the object.

The second index was the "reconstruction of an invisible whole from a visible fraction." They successfully recovered an object that had been partially covered by a screen while they were looking at it, that is, when only a portion of it was still visible. This was accompanied, on the other hand,

TABLE 4.
Stage 3 of Sensorimotor development in Cebus

	Age in days		
	Ca	Ro	Pe
- Sequences of primary schemata applied to objects	59	70	74
- Secondary circular reactions	?	?	?
OBJECT CONCEPT			
- Visual accommodations to rapidly moving objects	59	63	74
- Recovering partially hidden objects	59	63	74
- Searching for hidden objects at the point of disap- pearance	73	84	88

by a loss of interest and no attempts to search when the object was instead totally covered by a screen.

Finally, toward the end of third month of age, subjects began to search even when the object was completely covered by the screen, but they went look and search for it in the hand of the experimenter which was holding the object immediately before its covering, i.e. to the point where they last saw it.

On the other hand, the second line of development was remarkably different than in the human child. The central behavior pattern through which Piaget defines stage 3, i.e., "secondary circular reactions", was instead practically absent from the behavioral repertoire of subjects at this stage.

All efforts to elicit secondary circular reactions were made, by presenting a wide variety of effect-producing objects (objects that would move or swing or rotate, objects producing sound or noise, or even light-flashes, all contingent on some simple action of the subject). The subjects noted the result of their action or that of the demonstration of the experimenter but never paid much attention to it and did not attempt to make it last or reproduce it.

On the other hand, all subjects showed the typical exploratory behavior of this stage, which consists in the sequential application of different schemata to the same object. All these schemata were and remained, however, primary: they would touch, smell, lift, regard, mouth, etc. the object.

Stage 4

The most important new capacity characterizing stage 4 is the coordination of independent schemata into a means-end relation. All three subjects began to display it during their fourth month of life. Then, many instances of such coordinations could be observed throughout the stage. Two of the most typical examples are reported in Table 5.

TABLE 5.
Stage 4 of Sensorimotor development in Cebus

	Age in days		
	Ca	Ro	Pe
MEANS-END COORDINATION OF SCHEMATA			
- Lifting a screen to recover a hidden object	107	97	102
- Pulling a string to obtain an attached object	125	97	102
OBJECT-CONCEPT			
- Recovering a hidden object by uncoordinated attempts	94	91	96
- Recovering a hidden object by lifting a screen and grasping the object	107	97	102
- Typical reaction	219	175	156

Object-concept, too, continued to develop, in this stage, along the usual course seen in children. When the object was hidden under a screen, the subjects did not look anymore into the experimenter's hand but always attempted to recover it. First, by manipulating the screen through undifferentiated and not well directed actions, i.e. rather than lifting or pulling the screen by scrambling on it. This activity, however, became more and more sustained and insistent (as if they "knew" that the object had to be there) and finally succeeded by multiplying the accidental displacements of the screen. Then, and at the same time when independent schemata became coordinated into a means-end relation, the hidden object was immediately recovered by a precise sequence of the two actions of lifting the screen and grasping the object.

On the other hand, none of the developments in the construction of physical causality, typically shown by children at this stage, was seen in cebus. The absence of secondary circular reactions noted in stage 3 persisted into this stage. Thus, coordinations of secondary circular reactions (like hitting an object against another which gets displaced), that are the means through which causal relations among objects themselves begin to be constructed, were also not manifested.

Finally, Ca at 219 days and Ro at 179 days were still showing the "typical reaction" of stage 4 object-concept.

Further developments

Contrary to what we have seen in the other two species of non-human primates tested, cebus did develop, though with an extreme delay with respect to children, at least some instances of what appeared to be secondary circular reactions. All three subjects, toward the end of their fourth month, began to show a schema consisting in lifting an object, banging it to the ground and sliding it on the ground toward themselves. At the beginning this schema was not repeated, but occurred in single instances. It was, however, applied to various objects. It eventually got differentiated into two separate schemata: that of banging an object to the ground and that of sliding it on the floor. As soon as they differentiated, both these schemata occurred in quick repetitions: an object would be banged several time in succession, or it would be rubbed back and forth on the ground. Finally, at 275 days in Ca, at 325 days in Ro, and at 254 days in Pe, clear and uncontroversial signs that the action was provoked and regulated by the effects that the object produced could be detected. Objects were banged or rubbed in bouts of three or four strikes then they were lifted and their lower surface was visually inspected, and then this cycle was repeated several times. Furthermore, the intensity and velocity of the banging and rubbing action, was often varied at each cycle, as well as the number of strikes.

Slightly later than this last development, instances of object-object combinations began to occur. Two object were put side-by-side, or one into the other, or even one on top of the other.

At this time, Ca also showed one clear instance of tertiary circular reaction. An object was grasped and thrown up into the air and then let fall to the ground. The action was repeated several times, systematically varying the height at which it was thrown, while at the same time both its trajectory in air and its hitting and rolling on the ground was carefully monitored: Ca always waited for the object to come to a still before grasping and throwing it again.

Finally, at 16 months, both Ca and Ro manipulated objects through a variety of secondary schemata, as it was found at the beginning of testing of logic development (Pe has not yet been tested), which will be presented in chapter 13 below.

DISCUSSION

Cebus rates of development were faster than human's, but at the same time much slower that those seen in macaques. They entered stage 2 only at the beginning of the second month. Stage 3 was reached at the beginning of the third month and stage 4 in the middle of the fourth. It should be added that, contrary to both macaque and gorilla, rates of development relative to children did not slow down across stages but rather accelerated, or, to say it in a more telling way, stage 1 relative duration was particularly long, as compared to stage 2 and 3. As the more formal comparison of the following chapter will argue, this fact seems to point to a relatively greater immaturity at birth than either macaque or gorilla.

Cebus development in the first two stages did not show any significant difference with that of children. Contrary to macaques and gorilla, where stage 2 patterns of prehension development appeared already different, cebus followed, with one exception, the same course of children. The only difference was that cebus did not develop prehension-sucking coordination before prehension-vision coordination. These two coordinations developed at the same time: attempts to bring to the mouth grasped objects began at the same time as attempts to grasp objects at sight, and at the same time they began to be successful.

On the other hand, cebus did not develop any pattern of "buccal prehension", as seen in the other species. This fact is a counterproof of the crucial role that the relative time of onset of independent locomotion plays in the development of prehension patterns. Contrary to macaques and gorilla, cebus were not capable of autonomously moving their body in space before being capable of grasping objects at sight (compare Table 2 and Table 3).

Hence, no earlier visual-locomotive coordination, enabling the animal to reach objects with its body before it can grasp them with its hand, gets established.

When cebus became capable of moving toward objects his hand prehension was already functioning. In the next stage, the animal was able to approach objects and move around them to explore them visually, something children cannot do, but did not directly mouth them: any other approach was based on the use of the hand. There was no competition, as seen in the other species, between these two patterns. Though not exclusive, as in children, exploratory behaviors based on the use of the hand were by far dominant.

Stage 3 development conformed to that seen in the other nonhuman primate species. The main feature of this development, the absence of "secondary circular reactions", occurred also in cebus. Object-concept developed through the same sequence as in children, as did the sequential coordinations of schemata applied to one and the same object, but these schemata were all primary. Stage 4 confirmed this course of development, as did in the other species. Object-concept achieved stage 4 object-permanence through the usual sequence and at the same time means-end coordinations of schemata made their appearance. On the other hand, coordinations of secondary schemata, involving object-to-object physical relations, were not manifested. Further development of the cebus, however, did show interesting differences with respect to the other nonhuman primate species. Though even in this species the systematic experimental behaviors of tertiary circular reactions and use of detached tool-objects typical of stage 5 development in children did not follow stage 4 behaviors, at least some instances of secondary circular reactions were manifested toward the end of the first year of life. They were, furthermore, accompanied by instances of object-object combinations, normally occurring, with a much wider variety and frequency, in children at stage 4 and 5. These behaviors became even more diversified at about 16 month of age, which is, however, a full year after our subjects achieved stage 4.

It seems as if at least some of the behavioral structures of sensorimotor intelligence that were always found to be absent in the development of all species of nonhuman primate tested (including cebus) did eventually witness a limited development in this species, but in an extremely delayed fashion. Possible reasons of this phenomenon will be discussed in the next chapter. Further tests designed to probe specific abilities of stage 5 and 6 were conducted on cebus and will be reported in the following chapters.

6 Systematic Comparison of Early Sensorimotor Development

Francesco Antinucci
Istituto di Psicologia, C.N.R.
Rome, Italy

METHOD

In this chapter we will compare the courses of development of sensorimotor intelligence of the nonhuman primate species whose longitudinal testing was reported in the preceding three chapters and of the human primate.

To enable some quantitative estimates, a few behavioral landmarks have been selected as defining achievement of developmental stages, though we are perfectly aware that sensorimotor development is a smooth and continuous process, with no abrupt transitions and hence no clearly marked boundaries between stages. However, since in this case we are interested in the relationship among different developments, and not in the definition of developmental stages, the inevitable arbitrariness involved in such a selection should be compensated by the operationality it offers.

The general criteria that have guided our specific selection of items (presence in all species, reliability of assessment, etc.) have already been stated in the longitudinal reports. We now add the criteria defining stage boundaries in terms of those items, and their justification.

Since development of schemata intercoordinations is the typical achievement of stage 2, we identify completion of stage 1 and beginning of stage 2 with the achievement of the first of these intercoordinations, that between hand and mouth, as evidenced by the capacity to bring directly (i.e., without tactual adjustment) the hand to the mouth. The corresponding item in our tables is labeled "Thumb sucking". For the same reasons, we identify completion of stage 2 and beginning of stage 3 by the completed development of these intercoordinations, as evidenced by the achievement of

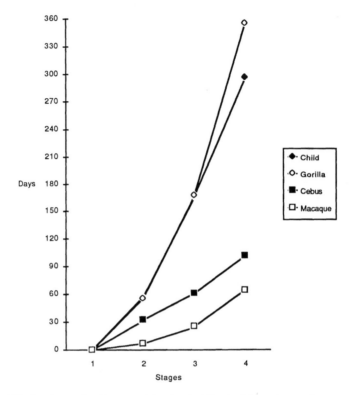

FIG. 1   Ages of achievement (in days) of the first four stages of sensorimotor intelligence development in each species.

the last one, i.e. that between prehension and vision. More precisely, the
boundary is marked by the achievement of perfect coordination between
vision and prehension: when the subject can grasp at sight objects in any
position (suspended objects being the most difficult) with a single precise
movement of the hand onto the target and no visual adjustment of the
hand. The corresponding item in the tables is "Fully developed visually
guided hand prehension".

Because of the significant interspecies differences in stages 3 and 4 (like
absence of "secondary circular reactions" in the nonhuman primate species), the boundary between these two stages has been identified on the
basis of the developmental sequence of object-concept. Since in this domain
the achievement of object permanence is typical of stage 4, we identify
completion of stage 3 and beginning of stage 4 with the appearance of the
capacity to recover an object which has been completely hidden under a
screen in full view of the subject. The action of recovering has to be
straightforwardly directed, without hesitations, and clearly differentiated

TABLE 1.
Age of Achievment (in days) of the First Four
Stages of Sensorimotor Intelligence by Each Species.

	Stage 1	Stage 2	Stage 3	Stage 4
Macaque	0	7	25	65
Cebus	0	32	61	102
Gorilla	0	56	168	356
Child	0	57	167	296

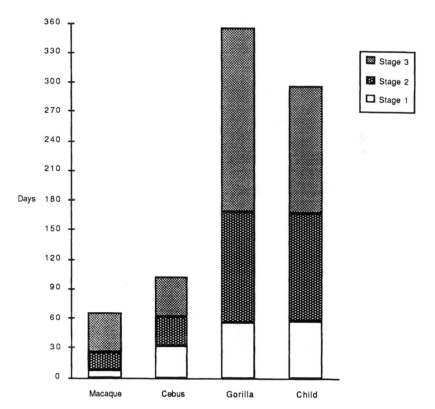

FIG. 2. Durations (in days) of the first three stages of sensorimotor intelligence development in each species.

into the two components of lifting the screen and grasping the object. This corresponds to the item "Recovering a hidden object by lifting a screen and grasping the object" in the tables.

In macaque and cebus the reported ages of achievement represent averages across all the subjects tested. In gorilla they are the ages of our single subject and in children the averages of Piaget's original subjects, as reported in Piaget (1971a, 1974).

RESULTS

Rates of Development

Table 1 and Fig. 1 show ages of achievement in days of the first four stages of sensorimotor intelligence development in the four species. Fig. 2 shows the same data as durations of stages.

Notice first of all the differences in absolute rates of development. Gorilla and child show the slowest rates: they are almost identical, the ratio between the total durations of their development from birth to stage 4 being .83. They are followed by the cebus, which is 2.90 times faster than the child, and then by the macaque, which is 4.55 times faster than the child. The data confirm a well-known fact: apes and humans class together as slow developers compared to monkeys. On the other hand, it is interesting to notice the difference between the two species of New World monkeys and Old World monkeys: cebus appears intermediate between ape and macaque.

Quite a different picture emerges, however, if one compares relative rates of development. Table 2 and Fig. 3 show, for each species, achievement of stages as a proportion over the total duration of development from birth to stage 4.

Both stage 2 and stage 3 are reached relatively earlier by macaque and gorilla than by the child. They are instead reached relatively later (especially stage 2) than the child by cebus. Fig. 4 shows the same data as relative duration of each stage for each species.

The macaque has the most accelerated development in both absolute (lowest curve in Fig. 1) and relative terms (lowest curve in Fig. 3): not only its development is the shortest of all species, but its first and second stages are also relatively shorter than in the other species. Gorilla, on the other hand, has the slowest absolute development (highest curve in Fig. 1), even longer than the child, but its stage 1 and stage 2 development is relatively quicker than in child, though much less so than in macaque. Cebus shows the opposite pattern: its development is quicker than both gorilla and child, though much slower than macaque, but its stage 1 and 2 development is relatively slower than in the child, in fact it is the slowest

TABLE 2.
Achievement of Sensorimotor Intelligence Stages
by Each Species Expressed as a Proportion of Total Development Between Birth and Stage 4.

	Stage 1	Stage 2	Stage 3	Stage 4
Macaque	0	10	.38	1
Cebus	0	.32	.60	1
Gorilla	0	16	.47	1
Child	0	19	.56	1

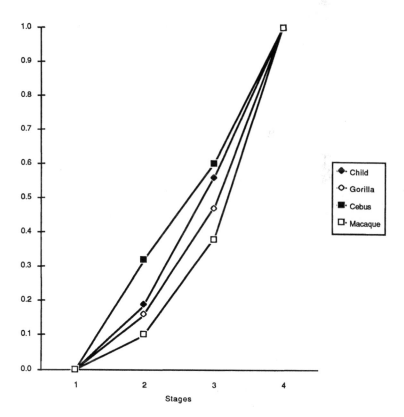

FIG. 3.   Times of achievement of each stage of sensorimotor intelligence
development relative to total time of development between birth and stage
four for each species.

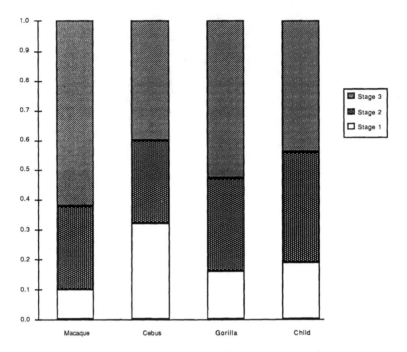

FIG. 4.   Relative durations of each stage of sensorimotor intelligence development for each species.

of all species (highest curve in Fig. 3). A different measure, taken this time across species, confirms this trend. If we split the overall developmental ratios of each species to the child, that were given above, in a stage by stage fashion and plot their courses, widely different profiles of the three nonhuman primate species emerge, as shown in Table 3 and Fig. 5.

The gorilla to child developmental ratio is close to 1 at all stages, indicating nearly identical developmental rates. It is, however, slightly higher in stage 1, but lower in stage 3, indicating that this species development progressively slows down with respect to the child, especially toward the end of the period considered. The macaque to child ratio starts very high, but it is nearly cut in half passing from stage 1 to stage 3, indicating a strong and also increasing slowing down of developmental rates with respect to the child. Cebus, on the other hand, shows an opposite trend: its developmental ratio to the child has a strong increase in passing from stage 1 to stage 2, and a slight increase from stage 2 to stage 3.

The first important lesson to be learned from this comparison is that one should be especially careful in equating developments of different absolute durations by applying some coefficient of proportionality, based on whatever measure (e.g., sexual maturity, tooth eruption, average life-

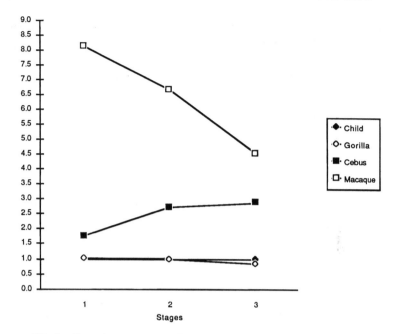

FIG. 5.   Stage-by-stage rates of development of each nonhuman primate species to the child.

span, etc.): the question is not finding the right measure, but the fact that any such measure is likely to give a misleading picture because development itself "develops", so to speak. If we choose achievement of object-permanence, for example, we would state that gorilla development is slightly slower than the child, cebus about three times faster, and macaque about four and a half faster. This statement, however, masquerades many more subtle developmental processes that are widely different from species to species, as we have just seen.

TABLE 3.
Stage-by-Stage Developmental Ratio of
Each Nonhuman Primate Species to the
Child.

	Stage 1	Stage 2	Stage 3
Macaque	8.14	6.68	4.55
Cebus	1.78	2.74	2.90
Gorilla	1.02	.99	.83
Child	1	1	1

In our specific case, there is a contrasting pattern of deceleration and acceleration within the different developmental rates, between macaque (and, to a lesser extent, gorilla), whose rates constantly decrease, on the one side, and cebus, whose rates increase, on the other side.

Though this difference is open to several interpretations, including the fact that deceleration and acceleration are only relative terms but do not say which species is doing what, a reasonable hypothesis might be advanced on the basis of the fact that where there is deceleration it tends to be strongest toward the end of the period of development considered, while the opposite is true of acceleration, which occurs practically only at the beginning (between stage 1 and stage 2, in cebus). Accordingly, it might be surmised that the first phenomenon reflects more the approaching of a limit, while the second something like a comparatively earlier start, i.e. a more pronounced relative immaturity at birth, thereby the first stage is proportionally longer.

Whatever the correct interpretation, however, this pattern is important to keep in mind, because, as we shall see below, a variable of the utmost importance in determining the eventual course of development is the temporal relationship among the relative times of onset of various capacities, and this relationship will be different in the various species because of their varying rates of development at each stage.

Stages of Development

Stage 1

In stage 1 capacities are least differentiated and consequently there is the highest degree of uniformity among species. Exercise of primary reflexes of grasping, sucking and visual fixation evolves into primary schemata for tactile, oral, visual (and, presumably, auditory) assimilation of external stimuli. These schemata are uncoordinated: the visual, oral and tactile worlds are separate and unconnected realities.

Stage 2

Stage 2 is marked by the establishment of reciprocal coordinations between primary schemata. In the human infant these coordinations develop in a characteristic sequence. The most primitive is the coordination of hand and mouth, usually manifested in thumb sucking. Then coordination between prehension and suction is established: objects that are grasped are taken to the mouth and viceversa. Finally the coordination between vision and prehension is established. Already at this point, the first important differences in the developmental courses of the four species appear. In the macaque hand-mouth coordination is well established at 6-7 days. However, no prehension-suction coordination follows. A completely different coordination, one that is never seen in the human infant, appears. Beginning at 7 days, on seeing an object the macaque grasps it directly with its mouth by targeting its head onto it (the pattern we labeled buccal prehension). This is possible because the macaque quickly develops a capacity for autonomous locomotion: at 7 days it can confidently move itself in space. Visually guided grasping through mouth requires a much more elementary coordination than visually guided hand grasping, because eyes and mouth are both carried on the head and, hence, simple visual targeting by means of locomotion brings automatically the mouth onto the object. During the second week of life buccal prehension becomes the dominant pattern of interaction with objects, while hand prehension guided by vision is still unsuccessful.

First partially successful attempts to visually guided hand prehension begin at 13–14 days, while the full coordination of vision and prehension is achieved at 25 days. Here the standard human sequence is followed: first prehension is successful only when both the hand and the object are in the same visual field, then the hand grasps without requiring any visual adjustment. Buccal prehension, however, does not disappear even when visually guided hand prehension is perfectly developed: it continues to occur side by side with hand prehension and is as frequent.

In the gorilla primitive hand-mouth coordination is well established at 56 days. At 66 days the first successful coordinations of prehension and mouthing occur, thus following the same sequence as in the human infant. Between 66 and 110 days attempts to grasp objects through visually guided prehension are all unsuccessful. In this period, however, there develops a pattern of buccal prehension identical to that seen in the macaque. In fact, at 88 days the gorilla becomes capable of independent locomotion in space and at the same time buccal prehension starts to occur and becomes the only pattern through which objects are picked up until 110 days.

At 110 days the first partially successful attempts to grasp objects through visually guided hand prehension occur and subsequently develop through the usual sequence. First the hand must be in the same visual field as the object and finally, at 168 days, objects are grasped with a single hand movement without visual adjustment.

During this phase buccal prehension continues to occur, but significantly decreases in frequency as hand prehension progresses. It appears that the two patterns are in a typical developmental competition: buccal prehension tends more and more to occur when hand prehension is unsuccessful because of either physical or cognitive difficulties, like grasping very small objects or partially hidden objects.

Cebus' development in stage 2 follows yet another course. Hand-mouth coordination is established very late at 32 days. First attempts to prehension-mouthing coordination begin to occur at the same time. However, these attempts are quite unsystematic and are also accompanied by attempts to grasp at sight. Neither of them is successful till 42 days: both developments appear to be very slow.

Contrary to macaque and gorilla, cebus does not develop during this period any buccal prehension schema. In fact, at this age the cebus is still not capable of independent locomotion. Like in children, only limited head movements are possible for it. Accordingly, some buccal grasping guided by sight is found, but only in the case of the nursing bottle when this comes close enough to its head.

At 42 days successful coordination of vision and prehension begins, and, as for the other species, only when both the hand and the object are in the same visual field. It is only at 44 days that the cebus begins to be able, by slowly creeping on the floor, to move freely in space. Prehension development is finally completed at 61 days, through the usual two steps.

Prehension and locomotion

Table 4 and Fig. 6 summarize the data on the acquisition of prehension and locomotion. To make comparisons more detailed, the development of prehension has been subdivided into 3 phases. Phase 1 corresponds, for all species, to the absence of any coordination. Phase 2 begins with the most primitive hand-mouth coordination and ends when grasping at sight begins to be successful. Phase 3 encompasses the development of visually guided hand prehension, up to its completion.

As we have seen, from Phase 2 the four species begin to differ from one another. The macaque begins this phase with a visual-locomotive coordination and a consequent buccal prehension pattern. No prehension-suction coordination ever appears, and the animal is capable of independent movements in space since the very beginning of this phase. Buccal prehension is the only prehension pattern of this phase.

The gorilla, on the other hand, develops in this phase the prehension-suction coordination, but, as soon as independent locomotion is established, the same visual-locomotive coordination of the macaque appears and buccal prehension becomes the almost exclusive grasping pattern till the end of this phase.

The cebus attempts in this phase both prehension-suction and vision-prehension coordination, but is unsuccessful until the end of this phase. Independent locomotion is absent through the whole phase and no visual-locomotive coordination appears. Buccal prehension is nonexistent.

The human infant develops in this phase only the prehension-suction coordination and, obviously, no locomotor ability.

TABLE 4.
Age of Achievement (in days) of the Three Phases of
Prehension Development and of Independent Locomo-
tion by Each Species.

	Phase 1	Phase 2	Phase 3	Locomotion
Macaque	7	14	25	7
Cebus	32	42	61	44
Gorilla	56	110	168	88
Child	57	157	167	255

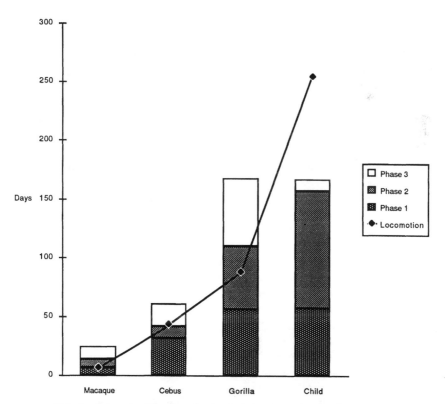

FIG. 6.   Ages of achievment (in days) of the three phases of prehension
development and of locomotion in each species.

Phase 3, that corresponds to the development of visually guided hand prehension, shows the same sequence for all species: successful grasping requires initially the visual adjustment of the hand to the object. Differences are present here too. The macaque continues to use buccal prehension side by side with hand prehension. In the gorilla the two prehension patterns compete with each other: buccal prehension becomes less and less frequent as hand prehension becomes more and more precise. The cebus acquires independent locomotion during this last phase but does not develop any subsequent pattern of buccal prehension. In the human infant, finally, motion is still limited to head-turning.

Stage 3

In stage 3 of sensorimotor development differences among the four species increase. In the human infant this stage is marked by the appearance and development of "secondary circular reactions", i.e., circular reactions where an action schema applied to an object produces an "interesting" result on the external environment, a result that captures the infant's attention and that he tries to conserve and reproduce by repeating the schema. This crucial behavior pattern was completely absent in the macaque: objects were always assimilated through primary schemata (mouthing, touching, lifting, smelling, etc.), even when actions on them produced salient effects that were clearly noticed by the animal (making noise, flashing light, swinging, rotating, etc.). Secondary schemata were never encountered, even much later in development.

Secondary circular reactions were also completely absent in the third stage of the gorilla's development. Some very primitive instances of this behavior pattern seemed to show up at the beginning of the second year of age, but they did not undergo any qualitative or quantitative development.

No secondary circular reactions were manifested in this stage by cebus either. Cebus, however, did begin to develop, in their fourth month of life, schemata that slowly evolved into full-fledged secondary circular reactions between their ninth and twelfth month. Though always limited in their variety and occurrence with respect to children, these schemata become more and more frequent and numerous in the first half of the second year of development.

All species, on the other hand, showed the beginning of the sequential coordination of independent schemata that is also typical of this stage. This appears in the so-called "object-exploration", where different schemata are applied one after the other to the same object.

The contrast between these two results (absence of secondary circular reactions, but presence of sequential schemata coordinations) is all the more interesting because secondary circular reactions are themselves a simple form of sequential schemata coordination: that between the action on an object and the subsequent effect that the object produces, the coordination being evident in their "circularity", i.e., in the attempts to reproduce the second by repeating the first.

Thus, it is only a specific type of coordinations which is absent in the nonhuman primate species: the coordinations where the object is assimilated by the schemata as an intermediary in a cause-and-effect relationship, rather than as a collection of physical properties, as it happens in the sequential coordinations of primary schemata.

These, however, are the coordinations that begin to construct the notion of physical causality. The development of object-concept is, instead, completely uniform in the four species. The same steps are followed, from adjustment to rapid movements, that marks the beginning of the stage, to the reconstruction of an invisible whole from a visible fraction, to the successful recovery of a completely hidden object, that marks the passage to stage 4.

Stage 4

Stage 4 confirms and amplifies the pattern of differential development seen in the preceding stage. Sequential coordinations of schemata evolve into the more complex, hierarchical means-end coordinations in all four species.

In the nonhuman primates, however, schemata thus coordinated are and remain all primary: they all involve subject-object relations, such as setting aside an obstacle and reaching an object, raising a screen and picking up an object, or pulling a string and grasping an object attached to it. Coordinations of secondary schemata, which realize object-object interactions, were not seen: one object was not hit against another to displace it, or pushed against another, or thrown against another, or stopped by means of another when in motion. It should be noticed that also schemata involving in general object-object relations, like putting one object on top of or inside another, or however combining one object with another, were absent in the nonhuman primate species.

These species, however, differed also from one another. In macaque even later development did not show any progress in this domain. In the gorilla one or two very primitive and elementary schemata involving object-object interactions (like banging) were found toward the end of the first year, but they did not seem to increase or progress. In cebus, too, schemata relating object-to-object are initially absent. But they eventually make their appearance, after quite a temporal lag: toward the end of the first year, for example, two objects are combined together side by side or on top of one another or one inside the other. Contrary to the macaque and the gorilla, however, at about 15–16 months there is a real explosion of object-object composing and provoked object-object dynamical interactions. Cebus, and only it, even showed in this period some instances of tertiary circular reaction.

Development of object-concept in this stage remains constant across the four species and it should be added that all nonhuman primate species eventually achieve also stage 5 of the object-concept development, though with quite a delay relatively to the child (see below, chap. 7).

DISCUSSION

Species differences found in the development of the reciprocal coordinations of stage 2 seem to have an obvious correlate: the relative time of onset of independent locomotion. And here the word "relative" needs to be stressed: it is the time of onset of locomotion in relation to the time of prehension development in each species that is crucial.

Fig. 7 displays this relation as a locomotion/prehension ratio for each species. Taking as a reference point the complete development of prehension, at the end of Phase 3, one can see that the macaque is capable of moving in space at about a quarter (.28) of its prehension development course, the gorilla at about half (.52) and the cebus at about three quarters (.72). The human infant, on the other hand, begins to be able to move long after its prehension development is completed (1.53).

These differences explain the different developmental patterns of the four species. The macaque can move toward objects before any hand coordination relative to object-grasping develops (beginning of Phase 2); hence, in order to mouth them, it can take itself to the objects. When hand coordinations develop this pattern of object exploration is already well established and functional and will not be substituted.

The gorilla develops locomotion a bit later, when its first hand coordination is already established, but before being capable of grasping at sight (middle of Phase 2). Hence here too objects to be explored are approached directly by the mouth and not through the hand. Since, however, visually guided hand prehension starts to develop shortly after, buccal prehension enters in competition with it and comes to be used more as a substitute pattern. The cebus, finally, develops locomotion when visually guided hand prehension is already functional (Phase 3) and consequently no previous pattern of direct mouth exploration could be established: object exploration is channeled through hand grasping and no pattern of buccal prehension ever develops. In this respect, it is the cebus pattern that comes closest to the human infant, as it can be seen in Fig. 7.

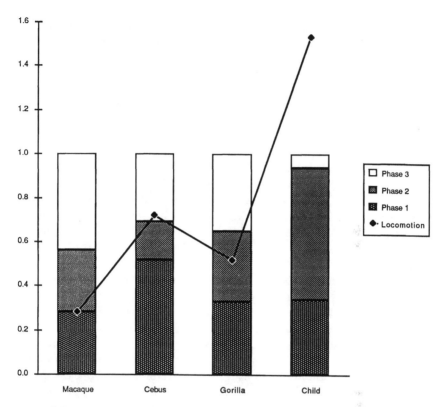

FIG. 7   Locomotion/prehension ratio in each species.

These early differences have consequences on the course of subsequent development. To understand why and how consider first what happens in child development. The human infant is practically immobilized for the first 9 months of his life. All his physical interactions with external objects, throughout this long period of his cognitive development, are mediated by the hand. Both his mouth and his visual exploration of objects depend on his hand grasping: since neither his body nor his head can move, it is the hand that must take objects to his mouth and it is the hand that can vary the position and perspective of objects (through movement and rotation) in order to allow visual exploration.

In addition to that, and again because of his immobility, the child has no way of interacting with distant objects (i.e., with objects outside of the range covered by his arm length) except through the use of other intermediary objects. If we consider that the hand is, in an elementary sense, the first "detached" instrument (and, in fact, it takes months to learn to move it under control and to direct it appropriately onto sighted objects), we can see that the child's exploration of and interaction with the environment, from the beginning of his first effective hand coordinations in stage 2, to the attainment of independent locomotion, which is well into stage 4, will develop almost entirely through the use of intermediaries: the hand at the more primitive levels, (the hand plus) other objects at the more advanced ones.

Consider now, at the opposite extreme, the macaque. At the very beginning of stage 2, it can move freely in space. It is still not capable of grasping objects with its hands, but it can approach them easily, it can mouth them, it can explore them visually by moving its body around them, it can climb on or even run after them. Furthermore, it can reach and explore also distant objects in exactly the same fashion, without having to find a way of acting on them indirectly. In other words, its exploration of and interaction with objects can occur without the use of intermediaries, i.e., through the simpler route of direct body action.

The explosion of mouth prehension, its dominance and concurrence with respect to hand prehension, the continued use of only primary schemata in exploring objects, all testify that this route is indeed taken.

The relative difference in the time of onset of locomotion channels very soon the cognitive interaction with the external world into two different modes: action through intermediaries in the child, direct action through body movements in the macaque. But action through intermediaries crucially depends on, and, hence, develops and exercises, control over shorter or longer cause-and-effect chains between the initial action and the final result, which in turn depend on and, hence, concentrate attention to the spatial and dynamical relations among the intervening objects.

This is the domain of development which begins with secondary circular reactions, where effects produced by an object acted upon are explored, continues with secondary schemata coordinations, where cause and effect interactions among objects themselves are explored, and schemata relating object to object, and culminates in tertiary circular reactions, where covariation between cause and effect is systematically experimented, and "instrumental" behaviors. And this is exactly the domain where no development occurs in macaque, even if, as we have seen, the general capacity for coordinating schemata sequentially and then hierarchically into means-end does develop as in children. In fact, development of object-concept which also depends on these coordinations but not on secondary schemata follows the same course as in children.

Gorilla and cebus fall in between these two extremes in terms of their relative times of onset of locomotion, but the gorilla shares more of the macaque pattern. In fact, as the macaque, it develops locomotion in Phase 2, that is, before the beginning of its hand-sight coordination, and hence the buccal pattern of prehension. This, however, regresses with the development of hand prehension, tough it does never disappear entirely. Consequently the gorilla's later developmental stages show essentially the same characters as the macaque's, though to a less extreme degree: limited and primitive secondary circular reactions and object-object relations seem to appear much later, and, especially, do not show any further development. In cebus, on the other hand, locomotion begins in Phase 3, and consequently no buccal prehension pattern ever develops. Its interaction with objects tends to occur through the use of the hand much more than in the first two species. Yet cebus too, contrary to children, is capable of approaching object directly through locomotion in stage 3 and 4, i.e. when schemata coordinations develop.

Accordingly, though with significant differences, cebus development appears as the most similar to the human one. Secondary circular reactions, coordinations of secondary schemata, object-object relation schemata are, in fact, found in this species, but neither in stage 3 nor for a good length of stage 4; they appear rather later, develop much more slowly and are less elaborated, differentiated and numerous than in the human infant.

The comparative minor scope of these structures is probably due to the fact that the action-through-intermediaries mode is less strong and exclusive than in children development. Furthermore, contrary to the gorilla, cebus does also show a slow but progressive development of these structures. This difference might be related to the different trends in the evolution of developmental rates seen above: while the gorilla rates progressively slow down as if approaching a developmental plateau, cebus rates do not slow down in the same period, but, if anything, tend to increase.

The second major difference found in the development of these four primate species, that of absolute developmental rates, might be important in accounting for differences of a more "quantitative" nature, rather than in the "quality" of the structures built.

Cognitive constructions develop through the subject's assimilative-accommodatory interaction with the environment. The rate of development directly affects the duration of this interaction at each successive developmental stage. One should consider, in fact, that the duration of each relevant environmental event (such as, for example, instances of specific spatial configurations, or object movements) or action schema acting on the environment, is practically the same for any developing subject. Faster developmental rates mean that the subject will experience less of these events and will generate less of these action schemata at each developmental stage. The correlate of this longer or shorter duration of the subject's interaction might well be a greater or smaller variety and richness of the constructions built at each stage. Within a given style of interaction, richer elaborations can be achieved depending on the amount of time spent in the interaction. We have seen that two different interacting modes characterize the macaque's and gorilla's developments on the one side, and the cebus' and child's on the other, but we have also noticed differences between the two species within each mode. The more quantitative of these differences might well be related to the longer or shorter "practice" allowed by the two species' different "times of exposure", determined by their vastly different absolute developmental rates.

CONCLUSION

A few general points emerge from this comparison. Major differences characterize the course of sensorimotor intelligence development of the human vs. the nonhuman primates. These differences start early and neither reduce to a slower or quicker progression of the stages nor amount to a radical departure from the standard stage sequence. Beginning from the primitive elaboration of the notion of physical causality of secondary circular reactions and ending with the study of causal and dynamical interactions among objects of tertiary circular reactions, they do show a coherent pattern. It is the fundamental cognitive domain of physical knowledge that is not being constructed by the nonhuman primate, or, at least, not to the same extent, as the human primate. On the other hand, logical knowledge, which in the sensorimotor period finds its origin in the structures of schemata coordinations (what Piaget calls the "logic of schemata"), seems to develop, at least up until the fourth stage in much the same fashion in the four species.

This difference appears to originate from the temporal interaction of prehension and locomotion development, that channels early cognitive interaction with the external world into different modes, which in turn "snowball," for the very effect of development, into larger and larger differences in later structures.

Insofar as this analysis proves to be correct, it sheds some light on possible paths of evolutionary transformation. Two "time regulators" appear to be involved. On the one side, rates of development (possibly tied to general rates of maturation) seem to have progressively slowed down, and in this case a near perfect concordance exists at the slowest level between ape and man.

On the other side, the temporal onset of locomotive capacity seem to have varied (possibly in relation to the different postural and locomotive adaptations of the four species) independently from the first factor. In this case the maximal difference singles out the human child with respect to the three other primate species. Interestingly enough, however, it is the South American cebus that is closer to man on this second dimension, and, at least in the domains we have considered, this species cognition appears to be more similar to that of the human child.

A more theoretical result of our comparison is that the sensorimotor stage, in the classical piagetian sense, and contrary to what Piaget himself seems to imply, does not appear to be a unit of cognitive structure. There is no necessary relation between secondary circular reactions and "intermediate" schema-detaching, as seen in object-concept, or between means-end coordination and the construction of object-object relations. On the other hand, it is the notion of domain of knowledge that seems to fulfill this role and have a unitary biological basis. The following chapters will probe more systematically the extent of the structural asymmetry between the two domains of logic and physics and its variations in the different species, as well as the extent and limits of their eventual development within each species. Chapters 9 and 10 will examine the understanding of physical causality and some elementary physical "laws," while chapters 13 and 14 will attempt to determine the level of organization of logical structures. Chapter 12 will again compare logical and physical components, as appearing in the spontaneous manipulations of objects.

It is important to stress that all the structures that will be investigated are at the sensorimotor level. In fact, even in the longitudinal studies, no attempt was made to probe the beginning or development of representational cognition: methodological difficulties in identifying behaviors qualifying for each species as "imitation" or "symbolic play," i.e. Piaget's precursors of true representation, are quite serious. Only in one field rather uncontroversial testing of a possible representational capacity could be confidently made, and it required, as we shall see, a deal more ingenuity than what is commonly required in testing children. Chapter 8 will examine whether the further development of object-concept attains a representational level.

III COGNITIVE DOMAINS

7
Stage 5 Object-Concept

Francesco Natale
Istituto di Psicologia, C.N.R.
Rome, Italy

INTRODUCTION

In stage 4 of sensorimotor development children, as well as the monkeys and ape we tested (see chapters 3, 4 and 5), become capable to recover objects that have been completely covered by a screen, provided they have seen them disappear under the screen. At this time it would seem that the object has become objectively and spatially permanent: it is correctly situated in space in relation to other objects and its existence does not depend anymore, as in previous stages, on the subject's actual actions and perceptions. Piaget, however, showed that in the human infant this is not yet the case: in fact, if an object is hidden under a screen and correctly recovered by the subject, but immediately after it is hidden under a second, different screen, still in full view of the subject, the infant will go search for it under the first screen, i.e., the place where it was hidden and successfully recovered the first time. Piaget termed this striking behavior pattern the "typical reaction" of stage 4. It shows that:

> The object is not a thing which is displaced and is independent of those displacements; it is a reality at disposal in a certain context, itself related to a certain action . . . the object remains dependent on its context and not isolated in the capacity of a moving body endowed with permanence. (Piaget, 1971a; p. 72)

The object is still partially tied to the subject's action: the infant searches for it in the place where his previous action was successful rather then in the place where he saw it disappear. Overcoming this "typical reaction" marks the beginning of stage 5 of object-concept development. In testing for stage 5 object-concept, however, care must be exerted, since when the child seems to have overcome the typical reaction, he may fall back into it when the task of recovering the object the second time presents some additional difficulties, for example, when the object undergoes a series of successive displacements or when the initially correct action does not succeed immediately. Piaget termed this second behavior "residual reaction". The testing procedure should, therefore, control for the existence of possible residual reactions.

Previous Studies

Object-concept and its development (most frequently called "object-permanence") is probably the most studied Piagetian cognitive category in the literature about non-human primate cognition. In particular, the attainment of stage 5 has been tested by Vaughter, Smotherman, and Ordy (1972) in squirrel monkeys, by Wise, Wise, and Zimmerman (1974) in rhesus monkey, by Mathieau, Bouchard, Granger, and Herscovitch (1976) in chimpanzee, tufted capuchin and woolly monkey, by Parker (1977) in stumptail macaque, by Redshaw (1978) in lowland gorilla, by Snyder, Birchette, and Achenbach (1978) in gibbon, tufted capuchin and rhesus monkey, by Wood, Moriarty, Gardner, and Gardner (1980) and Mathieau and Bergeron (1981) in chimpanzee.

In spite of the wide variability of the testing procedures used in these studies, ranging from the Wisconsin General Testing Apparatus (WGTA) to the Uzgiris and Hunt (1975) scale, they all agree in finding successful attainment of stage 5 object-concept in all species tested.

Yet an evaluation of their results raises some problems. The most important of these is that of distinguishing between learned, task-induced responses and responses implying a real understanding of the spatial position of the object. The first might easily occur when the testing procedure is a standardization of some kind of displacement task repeated with no or few variations over a large number of trials. In this case the subject might develop some task specific strategy, based, for example, on the absolute or relative position of the screens, or on a rule of sequentially uncovering the screens, etc. (see Antinucci, 1981; Thomas & Walden, 1985).

One way of protecting oneself against this danger is to vary, as much as compatible with due requirements of standardization, the context and the materials of the task, as Piaget himself always did, in order to prevent as much as possible the development of habits specific to the task. Nevertheless the majority of the studies quoted above failed to adopt any such protection.

In Mathieau et al. (1976), for example, the boxes used for hiding the object were always the same and never changed their relative position. Furthermore, an operant conditioning technique was adopted in order to familiarize the subjects with the testing apparatus. Even if an analysis of error distribution is missing, we can infer from the authors' discussion that errors in *Lagothrica* occur on first trials, which is a strong indication that some specific learning set is operating. Furthermore, there are no developmental data on the subjects tested and thus the differences among species found in the study might well reflect differences in learning abilities rather than a different level of object concept development. Similar objections can be addressed to Vaughter et al.'s study of object permanence in *Saimiri sciurea*. Using a modified version of WGTA, the authors tested three subjects, respectively 6, 9 and 12 months old, which were trained, through an extensive pretest, to uncover a cube when a reward was hidden under it. While the 6 month old subject did not solve the task, the 12 month old did; the 9 months old passed from 20% to 80% of correct responses during the course of five testing sessions. Again a sign that a task induced learning was occurring. This hypothesis is further confirmed by a so-called "post test", in which both the 9 and the 12 month old subjects went on uncovering the cube even when they saw that no reward had been under it. In spite of the authors' (fanciful) interpretation that this response is a sign of stage 6 level, it appears more probable that the older subjects simply acquired, more or less quickly, the behavior pattern of uncovering the cube when they fronted the experimental apparatus. WGTA was also used by Snyder et al. (1978) in testing gibbon, tufted capuchin and rhesus monkey. In this case the problem consists in the complete absence of any specification of the testing procedures used for evaluating object permanence. The authors simply declare: "Three Piagetian tasks for object permanence were administered . . . All subjects significantly demonstrated object permanence." (Snyder et al., 1978, pp. 946–947).

Of course, what is being questioned here is not the attainment of stage 5 object concept in gibbons, or cebus, or saimiri, or macaque, etc., but only the fact that studies like the ones discussed above cannot give any information on this topic.

Somewhat different problems raise studies that adopted the Uzgiris and Hunt or similar ordinal scales (Redshaw, 1978; Wood et al., 1980; Mathieau & Bergeron, 1981). Such scales have been developed and validated by testing large populations of human infants. The items chosen, the number of presentations required, the scoring criteria adopted are the result of an empirical selection process that tries to minimize the testing effort by eliciting only those behavioral responses that in the course of this process have been found to be crucial to assess reliably the presence of a given capacity.

Now, given the different behavioral repertoire displayed by different species of primates, there is no guarantee that these same items will elicit responses that are crucial in the same sense in a different species. In fact, our experience in testing longitudinally cognitive development in four species of nonhuman primates suggests that, had we used only these scales to assess the first four stages of sensorimotor development, we would have been seriously mislead in our evaluations. Furthermore, in some of the studies quoted above (Redshaw, 1978; Mathieau & Bergeron, 1981) this intrinsic weakness is enlarged by the complete absence of any quantitative data on the animals' behavior: not only any reference to typical or residual reaction or any error analysis is missing, but also no information is provided about the number of trials administered and the threshold value adopted to score a performance as correct.

Finally, we should add that another way to minimize the risk of misleading interpretations consists in integrating any experimental data obtained about object concept with data relative to other aspects of cognitive development of the subjects tested. Subjects followed longitudinally and tested on several capacities allow a more reliable understanding of the behavior patterns they exhibit, as studies by Parker (1977) and Wise, Wise and Zimmermann (1974) seem to confirm. In the present study of stage 5 object concept in nonhuman primates all the precautions discussed above were taken. In fact, four of the subjects tested (gorilla, Japanese macaque, and two cebus monkeys) were the same that had been followed longitudinally since birth and whose cognitive development in the first four sensorimotor stages is described in chapters 3, 4 and 5. To these three more subjects, all crab eating macaques, were added. Though these had not been followed since birth, another subject of the same species, which could not be tested, had.

METHOD

Subjects

Subjects were the gorilla (*Gorilla gorilla gorilla*) female Rm, at 18 months of age, (Rm18), the Japanese macaque (*Macaca fuscata*) female Oi, also at 18 months (Oi18), three crab eating macaques (*Macaca fascicularis*), two males and one female, aged between 26 and 28 months (respectively Pa26, Pi28 and Ch27), which were born in captivity and housed together in our lab, and the two cebus (*Cebus apella*) females Ca and Ro, at 12 and 7 months of age, respectively (Ca12 and Ro7).

Apparatus

The apparatus consisted of a wooden board (70 × 70 cm) and a set of 12 empty plastic blocks, all of different color and size (side length varying from 4 to 9 cms.), used as screens. A subset of 4 (or 5, depending on the experimental conditions described below) blocks were randomly selected for each session, out of which 2 (or 3) blocks were taken for each trial, so that the same block could not occupy the same position on two successive trials. The reward was a small (diameter = 1 cm) round flat candy or a seed of peanut. The wooden board, with the blocks aligned in front of the subject, was located at the level of cage floor for the gorilla, Japanese macaque and cebus, and at the level of perches internal to the cage for the crab eating macaques, since these animals usually do not stay on the floor.

Procedure

Two different kinds of displacement, called "between trials displacement" and "within trial displacement", each with two levels of complexity, were utilized. The four conditions given are diagrammed in Fig. 1.

In "between trials displacements" ("a" and "b" conditions in Fig. 1) the object was hidden under a block in full view of the subject, the apparatus was moved to within reach of the subject and the subject was allowed to recover it, then the apparatus was withdrawn and the object was hidden again under the same or a different block. At the first level of complexity (a) there were two blocks on the board, at the second level (b) there were three blocks; since trials of type (a) proved to be uninformative on the first three subjects tested, they were not administered to the three crab eating macaques and to the second cebus (Ca12). In "within trial displacements" ("c" and "d" conditions in Figure 1) the reward was hidden under a block and then moved to under a second block before allowing the subject to recover it. In the simpler condition (c) with two blocks there was one displacement, while in the condition with three blocks (d) there were two successive displacements before allowing the subject to respond.

If the animal turned over the block where the reward was, and only in that case, the trial was scored as correct; if it turned over any other block, either before or after turning over the right one, the trial was scored as incorrect.

Tests were administered by two experimenters: one manipulated the blocks and the other watched the animal to see if it was following all the steps of the procedure. Trials in which the animal did not pay attention to all the steps were interrupted and discarded. In order to secure maximum attention and best performance from each animal sessions did not have a

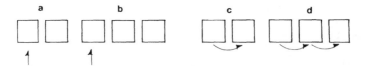

FIG. 1. Diagram of the four conditions used in testing for stage 5 object-concept. Straight arrows indicate that the reward was put under one of the blocks but not displaced ("a" and "b" conditions). Bent arrows indicate that the reward was successively displaced (once in "c" and twice in "d") from under one block to under another before allowing retrieval.

fixed, predetermined number of trials. Their length was determined each time "on line" by the attention and interest span of the animal.

RESULTS AND DISCUSSION

Tables 1 and 2 show results relative to, respectively, "between trials displacements" and "within trials displacements". As it can be seen, all subjects show overwhelmingly correct responses to all testing conditions: even in the most complex task, (d), errors never exceed the confidence limit. The lack of any consistent drop in performance across the four conditions and the variations introduced into each of them make it difficult to hypothesize the occurrence of any task specific learning.

Errors were analyzed to determine how many of them were residual reactions. To do this, we considered all couples of consecutive trials and calculated in how many of them a residual reaction could have occurred. Columns marked R.R. in both Table 1 and 2 report absolute frequencies of residual reactions as a proportion over the total number of possible occurrences.

TABLE 1.
Responses to the Two Conditions of the "Between Trials Displacement" Task.

Subjects	Two Blocks (a)			Three Blocks (b)		
	Correct	Incorrect	R.R.	Correct	Incorrect	R.R.
Gorilla(Rm18)	6*	0		6**	0	
M.fuscata(Oi18)	34**	4	1/20	11**	0	
M.fascic.(Pi28)	—	—		43**	7	0/19
M.fascic.(Pa26)	—	—		70**	2	0/54
M.fascic.(Ch27)	—	—		24**	4	2/11
Cebus (Ca12)	—	—		21**	0	
Cebus (Ro7)	22**	2	1/10	24**	0	

R.R.: Residual Reaction
*: Binomial Test, $p < 0.02$
**: Binomial Test, $p < 0.001$

TABLE 2.
Responses to the Two Conditions of the "Within Trials Displacement" Task.

	Single Displacement (c)			Double Displacement (d)		
Subjects	Correct	Incorrect	R.R.	Correct	Incorrect	R.R.
Gorilla(Rm18)	10*	2	2/8	13**	4	4/12
M.fuscata(Oi18)	20**	0		20**	0	
M.fascic.(Pi28)	40**	8	4/40	44**	4	1/40
M.fascic.(Pa26)	48**	0		48**	0	
M.fascic.(Ch27)	25**	2	1/21	42**	3	1/37
Cebus (Ca12)	12*	4	2/11	32**	0	
Cebus (Ro7)	47**	3	3/34	44**	3	3/32

R.R.: Residual Reaction
*: Binomial Test, $p < 0.05$
**: Binomial Test, $p < 0.001$

All errors of both gorilla and cebus Ro7 in the most complex task were due to residual reaction (see Table 2, rows 1 and 7), while in other subjects errors due to residual reaction were equal or less than other errors. Since gorilla and cebus Ro7 were the only subjects that were tested soon after the completion of stage 4 of sensorimotor development, this result could indicate that they were still in their transition to stage 5. Anyway, even the gorilla, which shows the highest proportion of errors in the most complex tasks, manifested a residual reaction only on 6 out of 20 possible occasions. There appears to be only traces of residual reaction: it occurs only in the most difficult conditions, and even in these cases there is no systematic falling back on it.

In summary, data relative to the frequency and nature of errors, and the complete absence of any pattern suggesting the occurrence of a task specific learning, converge in confirming the attainment of stage 5 in the development of object concept in all the species tested.

8 Stage 6 Object-Concept and Representation

Francesco Natale and Francesco Antinucci
Istituto di Psicologia, C.N.R.
Rome, Italy

INTRODUCTION

In the developmental sequence of sensorimotor intelligence, achievement of the last stage, Stage 6, marks a very important step in cognitive development. Stage 6, in fact, is characterized by the appearance of *representational* cognition in contrast to the *presentational* cognition typical of the first five stages of sensorimotor intelligence. In Piaget's own words:

> Due to representation "mental experience" succeeds actual experimentation . . . the subject represents the data offered to his sight otherwise than he perceives them directly. In his mind he corrects the things he looks at: that is to say, he evokes positions, displacements or perhaps even objects without actually contemplating them in his visual field. (Piaget, 1974, p. 351)

From this point on, cognitive action needs no more to be carried out through eyes and hands, as it was in the whole of the sensorimotor period, but becomes internalized: it might be carried out in the mind through mental actions on mental objects that stand for or "represent" reality (or its relevant features). In doing so, however, it enormously increases its power. Not only it becomes obviously much faster, but it might also work on possible worlds, rather than only on the actually presented world, and it might act in ways that are not possible in the actual world because of its physical constraints; for example, it might act backward, inverting the result of a positive action, even when that would not be possible in the actual world. Possibility and reversibility are two crucial dimensions added to its domain. Finally, the correspondence relation that representation creates is the essential prerequisite for symbolic functions.

In fact, it is through the appearence of the first rudimentary symbolic behaviors that Piaget tracks the origin of representation in children, in his work specifically dedicated to this topic (Piaget, 1951). As we said above, however, identifying infants' behaviors qualifying as "symbolic play" or "imitation" in each of the nonhuman primate species followed longitudinally poses serious problems, at least methodologically (just to mention the simplest one, some of our subjects did not have an adult conspecific living with them).

On the other hand, as we have seen, we could ascertain that all species tested reached stage 5 competence in the development of object-concept. Since in stage 6 the construction of object-concept becomes representational, one could test for the presence of representational capacities by trying to ascertain whether these species reach stage 6 competence in the development of object-concept.

Previous Studies

In the literature, successful achievment of stage 6 object-concept has been reported not only for apes (gorilla: Redshaw, 1978; chimpanzee: Wood, Moriarty, Gardner & Gardner, 1980; Mathieau & Bergeron, 1981;) but also for a variety of lower primate species (rhesus monkey: Wise, Wise, & Zimmermann, 1974; woolly monkey and cebus: Mathieau, Bouchard, Granger, & Herscovitch, 1976; squirrel monkey: Vaughter, Smotherman, & Ordy, 1972). It seems to us, however, that the testing procedures employed in these studies do not guarantee that the animals are making use of representation, even when their performance appears to be successful. All the studies referred to above employed some version of the "invisible displacement" task: The experimenter shows to the subject an object to be retrieved, hides it into a container, and then passes the container under or behind a series of screens. The target object is surreptitiously dropped off under or behind the last screen passed, and the experimenter exposes the empty container and allows the subject to act. Searching for the target object under or behind the correct screen is scored as a successful response.

First of all, it should be noticed that in these studies, successful performance in this single task is equated with the presence of representational capacity, though Piaget himself explicitly and repeatedly warned that solving the invisible displacement task does not by itself imply the use of representation (Piaget, 1971a). Redshaw (1978), Wood et al. (1980) and Mathieau and Bergeron (1981) all tested for stage 6 by means of the Uzgiris and Hunt scale (1975). However, Fischer and Jennings have convincingly argued that simply passing the relevant items of this scale does not provide sufficient evidence for the presence of representation:

> Looking first at the last screen touched may well require some sort of representation. Most of the studies before Corrigan's, however, did not test for this strategy in a convincing way. They typically did not control for the spatial position of the last screen or the effect of previous success in finding the hidden object at a particular location. In the simplest procedure used in previous studies, the adult always hid the object in the same direction, so that the child could solve the task by always looking in the same position, such as the end screen on the right (Uzgiris & Hunt, 1975, Scale 1, Task 14, Criterion d). Even with more complex hiding procedures, many studies confound "last place touched" with one or two hiding positions. A convincing study would show that the child searches in the last place touched, even when the following conditions are varied: the position of the last screen touched and the order in which the screens are touched. (Fischer & Jennings, 1981, pp. 20–21)

Results obtained by Vaughter et al. (1972) and Mathieau et al. (1976) are even less convincing, because the procedure employed is hardly comparable to the invisible displacement task. In their study, in fact, the target object was not hidden in a container at all. Its displacements under the screens were made invisible by holding another screen in front of the subject. In this way neither the action of displacing nor the path of the object could be seen.

Wise et al. (1974), on the other hand, used the standard invisible displacement task and systematically varied the position of the last screen touched and the order in which the screens were touched. However, as Fischer and Jennings (1981) noticed, though searching in the last place touched when this procedure is followed is "a reasonable candidate for a search behavior that requires representation" (p. 21), it is by no means a fully convincing one.

> A more convincing demonstration of the mental recreation of the invisible displacements of the object would be children's ability to reproduce the path of the hider's hand, a phenomenon that Corrigan, Uzgiris and Hunt, and others have called "systematic search." (Fischer & Jennings, 1981, p. 21)

Systematic search, however, has proved to be quite an elusive phenomenon, and many researchers have been unable to elicit this pattern (see Corrigan, 1981). In fact it should be recalled that Piaget himself noticed the occurrence of this behavior only once: "During test 7, Jacqueline *even* touches the three screens in succession, following the order in which I myself had slid in and withdrawn my closed hand . . ." (Piaget, 1971a, p. 89, italics added)

To sum up, it seems that in order to test for the presence of representation by means of the invisible displacement task, one should bear in mind the following three points: (a) systematic search cannot be the criterion because it cannot be elicited as a consistent behavior pattern; (b) going to the last screen touched cannot be, by itself, a sufficient criterion, even when conditions of hiding are systematically varied; (c) the invisible displacement task can be resolved in nonrepresentational ways or, as Piaget reported, ". . . merely through empirical or practical apprenticeship, in which case there would not be a true image of invisible displacements" (Piaget, 1971a, p. 78). It seems to us, therefore, that one should not only carefully vary and control all the parameters of the invisible displacement task but also employ additional tasks that can differentiate the representational or nonrepresentational nature of the responses given to the task. The two experiments reported here are an attempt to follow such a strategy in testing the attainment of stage 6 object concept in the four species of nonhuman primates, whose cognitive development has been reported in the preceding chapters: Japanese macaque (*Macaca fuscata*) and crab-eating macaque (*Macaca fascicularis*) (chapter 4), gorilla (*Gorilla gorilla gorilla*) (chapter 3), and cebus (*Cebus apella*) (chapter 5). The two experiments were performed separately and are also reported separately because the second was built and administered on the basis of the experience and results gained from the first. In the first experiment, whose subjects were the gorilla and the Japanese macaque, we were trying to overcome the many difficulties and uncertainties that plagued previous experiments, as discussed above. Consequently, the form of the task was progressively adjusted in order to disambiguate results that were being obtained. This led to the elaboration of a final satisfactory testing paradigm that could avoid most of the pitfalls encountered. This form was then employed in the second experiment, whose subjects were the crab-eating macaques and cebus.

Experiment 1

In order to fit the methodological requirements discussed above, we re-established the original piagetian version of the invisible displacement task: the container into which the target object is hidden and displaced is not exposed empty after the displacements, but left upside down. This was done for two reasons. First, because on the basis of Piaget's original data, we thought that this procedure could elicit more easily some pattern of systematic search: if the subject searches first where he has seen the object disappear and then under one of the screens, there is some evidence that he is retracing the path of the object and reconstructing its invisible transfer. Second, because this procedure allowed us to introduce into the experimental paradigm some "false trials" that, as it will be seen below, could reveal the real nature of the strategy of uncovering the last screen touched.

In addition, we modified in successive steps the invisible displacement task driven by the following logic: when successful performance in one version of the invisible displacement task was reached, which according to the standard testing procedure would indicate the presence of representation, some parameters of the task were modified in ways that would be irrelevant if the subjects were indeed solving the task by mentally representing the path and the invisible transfer of the target object, but would affect his performance if he was instead making use of some kind of practical search strategy.

METHOD

Subjects

Subjects were the gorilla female (Rm), at 22 months of age, and the Japanese macaque female (Oi), also at 22 months of age. Both subjects were tested after they gave evidence of having attained stage 5 of the object concept series (see chapter 7).

Apparatus

The apparatus consisted of a wooden board (70 × 70 cm), a set of three empty plastic blocks, open on one side, whose side-length ranged from 3.5 to 5 cm (small blocks), and a set of six similar blocks, whose side-length ranged from 7 to 9 cm (large blocks). The blocks were all of different colors and dimensions. Rewards were small (1 cm of diameter) round flat candies.

Procedure

On each trial, three blocks (one small and two large) were aligned on the wooden board with the open side down at a distance of 12 cm from each other.

In full view of the animal, a reward was placed under the small block. The small block was then passed under one of the large blocks by sliding it over the board. The large block was lifted from behind (relative to the animal's point of view) by tilting it along the front edge, and the reward was left under it. The small block was then moved back into alignment with the large blocks, always sliding it with the open side down. In this way, the animal could not see the transfer of the reward from the small

TABLE 1.
Responses to Test 1.

Subject	S	D	E	Total
Gorilla Rm22	4	20	0	24
Macaque Oi22	3	16	8	27

Note: S: Sequential responses; D: Direct responses; E: Errors.

to the large block. The board was then moved toward the cage bars so that the animal could easily reach the blocks by extending its arm through the bars. Three kinds of responses were scored: Sequential responses (S), the subject searched first under the small block where it had seen the reward disappear, and then under the large block where the reward had been left; Direct responses (D), the animal searched directly under the large block where the reward had been left; all other responses were scored as errors (E). Obviously, these responses are meaningless, relative to the presence or absence of representation, unless one is sure that the subject has been following the whole procedure of hiding and displacing the reward. To ensure this, the following precautions were taken:

(a) Tests were administered by two experimenters: one manipulated the blocks and the second one watched the animal to see if it was following all the steps of the procedure. Trials in which the animal did not pay attention to all the steps were interrupted and discarded.

(b) A test session was terminated as soon as the animal began to show signs of falling attention or lack of interest. Sessions, therefore, did not have a fixed and constant number of trials.

A series of five tests in sequence was administered. In Test 2, 3 and 5 the basic procedure described above was slightly modified (in ways that will be discussed below) in order to disambiguate the results obtained in the preceding test.

RESULTS AND DISCUSSION

Test 1

Test 1 was administered in two sessions. The same three blocks were utilized in all trials. Direction of movement of the small block (left-to-right, right-to-left) and large block lifted were alternated at each trial. Results are shown in Table 1.

Both animals gave a majority of direct responses: they searched directly under the large block, thus giving the typical response scored as correct

TABLE 2.
Responses to Test 2.

Subject	S	D	E	Total
Gorilla Rm22	27	8	4	39
Macaque Oi22	5	27	2	34

Note: S: Sequential responses; D: Direct responses; E: Errors.

and used as evidence of attainment of Stage 6 in studies employing the invisible displacement task. Though this kind of response might be indicative of a representational reconstruction of the itinerary of the reward, and in our case, the action simply skips the passage of searching under the small block, it might also be the result of a simple task-specific learning when trials are repeated: the subject simply learns that the reward will somehow be found under one of the large blocks, namely, the one manipulated by the experimenter, without reconstructing its invisible passage and transfer from the small block. In order to counter such a practical strategy, we tried to make it more difficult to identify the large blocks across trials.

Test 2

Test 2 was administered in three sessions. The same procedure as in Test 1 was adopted, except that all of the blocks were changed at each trial so that their color and size were different each time. Results are shown in Table 2.

There is a clear difference in the two subjects' performance (chi square = 25.8, $p < 0.001$). The gorilla gave a majority of sequential responses: it first searched under the small block where it had seen the object disappear and then, on realizing that the reward was not there, it went to search under the large block where the small block had passed. The consistency of this systematic search patterns when all components of the experimental situation vary at each trial might be evidence that the gorilla was indeed mentally reconstructing the path and the transfer of the target object. On the other hand, the macaque continued to give a majority of direct responses.

How can we interpret these results? The fact that the gorilla adopted the sequential search pattern only in Test 2, whereas it gave direct responses in Test 1, strongly suggests that the strategy of searching directly under the large block was not based on the representation of the invisible transfer, but rather on some practical rule like "go to the large block manipulated

TABLE 3.
Responses to False Trials in Test 3.

Subject	Correct	Incorrect	Total
Gorilla Rm22	12	0	12
Macaque Oi22	8	8	16

by the experimenter", a rule that could be more easily built when the two large blocks were physically the same and occupied the same positions across trials as in Test 1.

Test 3

To test this hypothesis, we introduced into the experimental paradigm a set of "false trials" interspersed with ordinary trials. False trials were trials in which, after hiding the reward under the small block, this was not moved to under a large block, but was left in its place. The experimenter then proceeded simply to lift one of the large blocks in the same fashion as he did when transferring the reward. If direct responses were performed on the basis of the practical rule hypothesized above, then one would expect the animal to search under the large block even when no contact at all had occurred between the small block containing the reward and the large block lifted, as in false trials, and hence no transfer could possibly have occurred. The ratio of ordinary to false trials was 2:1. Ordinary trials were as in Test 2. Two sessions were administered. The performance on false trials is shown in Table 3.

Out of 16 false trials, the macaque uncovered the small block where it had seen the reward being hidden only 50% of the times. The gorilla, on the other hand, never made an error in false trials. This fact seems to confirm that the gorilla was indeed solving the invisible displacement task by mentally reconstructing the unseen displacements of the target object. The macaque's results, on the contrary, clearly show that the invisible displacement task can be (and was, in fact) solved by means of a practical rule that does not have anything to do with representing the invisible displacements of the object.

At this point, the problem is the following: does the macaque rely on this practical, nonrepresentational strategy because it does not have the capacity for representation or simply because it does not need it, since in the ordinary invisible displacement task, as presented in Tests 1 and 2, the practical rule leads to a successful performance anyway? However, if the invisible displacement task includes false trials, as in Test 3, then this rule will not work and, hence, we might hope to push the macaque into using

TABLE 4.
Macaque's Responses to False and Ordinary Trials in Test 4.

Trial	Correct	Incorrect	S	D	E	Total
False	11	1				12
Ordinary			20	5	7	32

Note: S: Sequential responses; D: Direct responses; E: Errors.

a representational strategy, if it is capable of doing so. For this reason, we continued to give the same mixture of ordinary and false trials to the macaque.

Test 4

This test was administered to the macaque only. The same paradigm of Test 3 was utilized. There were six sessions with a total of 130 trials. During this test the macaque gradually learned to respond correctly to false trials. Table 4 shows the results of the last two sessions cumulated.

As it can be seen, the macaque's performance on false trials is now overwhelmingly correct (11 times out of 12). However, its response pattern on ordinary trials also changed: whereas direct responses were predominant before the introduction of false trials, now sequential responses are. Table 5 compares these responses with those obtained in Test 2, before the introduction of false trials: The inversion in frequency between the two kinds of responses (direct and sequential) is highly significant (chi square = 26.78, $p < 0.001$).

In other words, the macaque is now using the same search strategy as the gorilla, and, in fact, their performances are quite similar, as it can be easily seen by a comparison between row 1 of Table 2 (gorilla's responses on ordinary trials) and row 2 of Table 5 (macaque's performance after the introduction of false trials).

TABLE 5.
Macaque's Responses to Ordinary Trials Before
and After the Introduction of False Trials.

Trial	S	D	E	Total
Before false trials	5	27	2	34
After false trials	20	5	7	32

Note: S: Sequential responses; D: Direct responses; E: Errors.

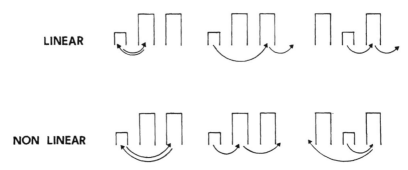

LINEAR

NON LINEAR

FIG. 1. Diagram of small block displacements in linear (top) and non linear (bottom) trials in test 5. Arrows trace movements of the small block.

One could conclude that the introduction of false trials disrupted the practical strategy utilized by the macaque, and forced it to pay attention to the actual displacements of the reward, because that could now be found both under the small block and under one of the large blocks. If this interpretation is correct, it means that the macaque is indeed capable of representing the invisible displacements of the object and had simply to be motivated to use such a capacity.

At this point, therefore, both animals were following a sequential search strategy. In order to be sure that this pattern was indeed generated by the reconstruction of the invisible path of the target object, we modified once more the experimental paradigm and gave both animals a further test.

Test 5

In order to meet the requirements of a more stringent criterion of testing through the invisible displacement task, as discussed in the introduction, in Tests 1–4 the direction of displacement of the small block and the itinerary it followed in passing under the large blocks were systematically varied at each trial. Yet, the small block always ended up in a position that was adjacent to the large block where the reward had been left, either on the right or on the left side of it (as exemplified in Figure 1, under the heading "linear" trials). In Test 5, trials were introduced in which the small block ended in a position that was separated from the rewarded large block by the second,untouched large block (as exemplified in Fig. 1, under the heading "nonlinear" trials).

The reason for this modification was that the first type of displacements allowed a solution by following a linear order in the uncovering of the blocks, whereas the second type of displacements, once the subject had uncovered the small block, required him to "skip" one of the large blocks in order to reach the correct one. If the animal is representatively reconstructing the invisible path of the reward, then this modification should

TABLE 6.
Responses to Linear Trials in Test 5.

Subject	Correct	Incorrect	Total
Gorilla(Rm22)	11	1	12
Macaque(Oi22)	16	1	17

not affect its performance, but if it is following some kind of practical rule on the basis of the relative position of the blocks, then its performance might be significantly altered. Four sessions were administered to the macaque and two sessions were administered to the gorilla. The ratio of linear to nonlinear trials was approximately 1:2. In Tables 6 and 7, an analysis of all the responses in which first the small block and then one of the large blocks was uncovered is given. Table 6 compares the performance of the two animals on the trials allowing a linear solution, whereas Table 7 compares their performance on the nonlinear ones. Correct and incorrect responses are, respectively, those where the large block uncovered was or was not the one that had been involved in the displacement.

Results on linear trials confirm the pattern already seen in previous test. Both animals were overwhelmingly correct in their responses, and their performance was not different (Fisher exact probability test, $p = 0.5$). Results on nonlinear trials, however, show a large difference in the animal's performance (Fisher exact probability test, $p = 0.019$). The gorilla still had a majority of correct responses: after uncovering the empty small block, it correctly skipped the adjacent large block to reach the one where the invisible exchange had occurred. The macaque's response pattern was quite different: after uncovering the small block, 29 times out of 43, it uncovered the large block adjacent to it, though this had not been interested by the exchange and, in fact, had not been touched at all by the experimenter. The macaque appears to have developed yet another kind of practical strategy: after searching under the small block, it sequentially goes on uncovering the large blocks.

Experiment 2

Two features of the tasks presented in Experiment 1 were especially useful in clarifying the nature of the subjects' responses. One was the introduction of false trials that enabled the disambiguation of direct responses. The

TABLE 7.
Responses to Nonlinear Trials in Test 5.

Subject	Correct	Incorrect	Total
Gorilla(Rm22)	13	8	21
Macaque(Oi22)	14	29	43

other was the introduction of nonlinear trials, which disrupted the success of a sequential search pattern (search pattern whose establishment, on the other hand, might have been facilitated by the initial presentation of only linear trials).

In Experiment 2 it was consequently decided to present right from start both linear and nonlinear displacements interspersed. As in Experiment 1, false trials would have been introduced if a prevalence of direct responses would have emerged.

METHOD

Subjects

Subjects were the three crab-eating macaques (*Macaca fascicularis*), already tested for stage 5 object concept (see chap. 7), aged between 27 and 29 months (respectively Pa27, Pi29 and Ch27), and the two cebus monkeys (*Cebus apella*), Ca and Ro, aged respectively 43 and 24 months.

Apparatus and Procedure

Both apparatus and procedure were identical to those described for Test 5 of Experiment 1, with the only differences that the linear to nonlinear trials ratio was 1:1. All possible combinations of positions and displacement of the blocks were used and balanced across trials. This lead to include also one more arrangement not used in Test 5: the one where, after the displacement, the small block ended between the two large ones. The basic form used in testing is diagrammed in Figure 2. Number of sessions administered varied from 8 (subject Ro24) to 12 (Pa27).

RESULTS AND DISCUSSION

Performances of the five subjects tested in Experiment 2 are reported in Table 8. As it can be seen, results are overwhelmingly negative for all subjects: they showed an overall number of errors ranging from a minimum of 43.1% (Ch27) to a maximum of 56.5% (Ca43). Error rates were even higher in nonlinear trials, where they ranged from 53.3% (Ch27) to 82% (Ca43).

No improvement in performance was found in going from the first to the last sessions, as it can be seen from the comparison between the error rate of the first half of the trials and that of the second half (see Table 9). If anything, the rate tends to increase in most subjects.

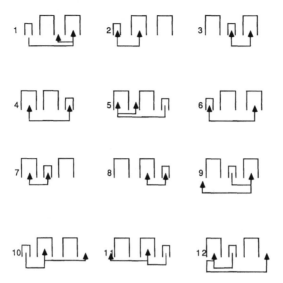

FIG. 2. Diagram of all small block displacements used in Experiment 2. Arrowed segments indicate the large block under which the reward is dropped and the final position assumed by the small block before allowing retrieval.

The only systematic tendency that emerges in this comparison is a decrease of sequential responses in favor of direct responses. A more detailed analysis of errors shows, however, that incorrect responses follow exactly the same tendency. As correct responses, errors, in fact, can be either "sequential" (uncovering the untouched large block after the small one) or "direct" (uncovering directly the untouched large block). Table 10 reports, for each animal tested, the relative weight of these two kinds of errors in the two halves of the experiment.

The picture that emerges is quite clear. At the beginning our subjects do search for the disappeared object first under the small block, that is where they have seen the object disappear, and then under one of the

TABLE 8.
Performances of the Four Subjects of Experiment 2.

Subject	S	D	E	Tot.
M. fascic. Pi29	20 (16.1%)	35 (28.2%)	69 (55.7%)	124
M. fascic. Pa27	26 (10.6%)	104 (42.4%)	115 (46.9%)	245
M. fascic. Ch27	26 (22.4%)	40 (34.5%)	49 (43.1%)	115
Cebus Ca43	24 (14.3%)	49 (29.2%)	95 (56.5%)	168
Cebus Ro24	15 (15.6%)	33 (34.4%)	48 (50.0%)	96

Note: S: Sequential responses; D: Direct responses; E: Errors.

TABLE 9.
Comparison Between Subjects' Performances in the First and in the Second Half of th
Experiment.

Subject		S	D	E	T
M.fascic.Pi29	I half	14 (24.1%)	15 (25.3%)	29 (51.8%)	!
	II half	6 (9.1%)	20 (30.3%)	40 (60.6%)	(
M.fascic.Pa27	I half	20 (15.6%)	49 (38.3%)	59 (46.1%)	1:
	II half	6 (5.1%)	55 (47.0%)	56 (47.9%)	1·
M.fascic.Ch27	I half	18 (25.7%)	22 (31.4%)	30 (42.9%)	:
	II half	8 (17.8%)	18 (40.0%)	19 (42.2%)	‹
Cebus Ca43	I half	17 (17.0%)	29 (29.0%)	54 (54.0%)	1(
	II half	7 (10.3%)	20 (29.4%)	41 (60.3%)	(
Cebus Ro24	I half	8 (16.7%)	15 (31.2%)	25 (52.1%)	‹
	II half	7 (14.6%)	18 (37.5%)	23 (47.9%)	‹

Note: S: Sequential responses; D: Direct responses; E: Errors.

large blocks. But their choice of the second block is random, as it can be seen in Table 11 that compares sequential correct responses and sequential errors in the first half of the experiment. The block that has been interested by the exchange and the one that has not even been touched are equally chosen: there appears to be no capacity of representing the invisible displacement.

Since they never find the reward under the small block and do not understand the invisible displacement, but, when accidentally successful, they always find it under one of the large blocks, along the course of trials our subjects shift strategy and begin to look directly under one of the large blocks. But since they do not represent the invisible displacement, they choose again randomly among the two of them, as shown in Table 12, where choices of the correct vs. the untouched, incorrect block are compared for the second half of the trials.

TABLE 10.
Comparison Between Sequential and Direct Errors in the Two Halves of the Experiment

Subject		Sequential error	Direct errors	Total
M.fascic.Pi29	I half	13 (44.8%)	16 (55.2%)	29
	II half	14 (35.0%)	26 (65.0%)	40
M.fascic.Pa27	I half	34 (57.6%)	25 (42.4%)	59
	II half	10 (8.6%)	45 (91.4%)	55
M.fascic.Ch27	I half	19 (63.3%)	11 (36.7%)	30
	II half	7 (35.0%)	13 (65.0%)	20
Cebus Ca43	I half	23 (42.6%)	31 (57.4%)	54
	II half	10 (24.4%)	31 (75.6%)	41
Cebus Ro24	I half	6 (24.0%)	19 (76.0%)	25
	II half	5 (21.7%)	18 (78.3%)	23

TABLE 11.
Comparison between correct and incorrect sequential responses in the first half of the Experiment.

	Sequential Responses	
	Correct	Incorrect
M.fascic.Pi29	14	13
M.fascic.Pa27	20	34
M.fascic.Ch27	18	19
Cebus Ca43	17	23
Cebus Ro24	8	6

CONCLUSION

Only the gorilla follows a systematic search pattern that consistently traces back the invisible displacements of the target object in quite a variety of different situations and, furthermore, its performance is not disrupted by the successive modifications introduced into the task. This evidence strongly suggests that this animal is representatively reconstructing the itinerary of the object and has thus reached Stage 6 in the development of object-concept.

On the other hand, Japanese macaque, crab-eating macaque and cebus seem to develop *ad hoc* practical rules, marked by different degrees of effectiveness, based on a variety of task specific cues. Thus they offer no evidence of a representational reconstruction of the invisible displacements of the target object and cannot be credited with the Stage 6 developmental level.

In view of the fact that all these animals fully developed Stage 5 object-concept (see chapter 7), the difference between the two groups seems to be entirely ascribable to the presence (gorilla) versus absence (macaques and cebus) of the capacity for representation.

TABLE 12.
Comparison Between Correct and Incorrect Direct Responses in the Second Half of the Experiment.

	Direct Responses	
	Correct	Incorrect
M.fascic.Pi29	20	26
M.fascic.Pa27	55	45
M.fascic.Ch27	18	13
Cebus Ca43	20	31
Cebus Ro24	18	18

This results suggests the existence of a split between nonhuman primates: insofar as representation is concerned, apes, but not monkeys, class with humans. Given the crucial role of representation in the development of any system of symbolic communication or symbol manipulation (Piaget, 1951), this result further suggests that monkeys, contrary to apes, lack the basic prerequisite for the development of any such system.

9 Causality I: The Support Problem

Giovanna Spinozzi and Patrizia Potí
Istituto di Psicologia, C.N.R.
Rome, Italy

INTRODUCTION

From the beginning of stage 3 of sensorimotor development fundamental differences have been found to characterize the development of nonhuman primate sensorimotor intelligence in comparison to that of the human infant. In chapter 6 we have shown as they seem to describe a coherent pattern: nonhuman primates appear to be selectively and strongly limited in the construction of the physical domain of cognition. The central pillar on which this construction rests and grows is the development of the notion of physical causality. In fact, it is the very absence of those behaviors that are specifically devoted to exploring cause and effect relations between objects (from secondary circular reactions to action through intermediaries) that, as we have seen, constitutes the most striking phenomenic difference with the behaviors observed in the developing child. In this and in the following chapter we will try to probe directly the nature and limits of the understanding of this notion in the nonhuman primate species whose cognitive development we have been following in the preceding chapters. The beginning development of this notion, in the third stage of sensorimotor development, is characterized by what Piaget calls "efficacy" or "magic phenomenalism":

> Causality through efficacy, characteristic of the third stage and the remains of which are observable well beyond the first seven or eight months of life, is a form of causal relation foreign to objective and spatial connection. When he acts or believes he acts on the external world, the child at the third stage

has no clear awareness of his body movements as objective displacements producing the effects perceived, and he has still less awareness of the intemediaries linking these body movements to the effects perceived. In the presence of an interesting sight which he has produced or which he desires to prolong, the baby reacts with a global attitude projected in differentiated movements, but the causal connection is not, for him, established between these movements, the series of intermediaries, and the final result; it merely links the global attitude, above all experienced from within, and the effect produced. (Piaget, 1971a, pp. 289–290)

It is precisely in the establishment of this "objective and spatial connection" as an essential condition for the existence of a causal relation that lies the fundamental achievement of the next two stages, that is, when "the child conceives spatial contact as necessary for the causal action of one body on another" (Piaget, 1971a, p. 292).

There are two indexes that reveal and attest the beginning of fully spatialized and objective causality. One is that the child's action will specifically attempt to establish physical contact with and between intermediaries: "If numerous traces of causality through efficacy linger until the end of the first year, it is nevertheless obvious that from about 0;9 the child comes to act upon things . . . through physical contact, pressures, attempts to set things in motion, etc." (Piaget, 1971a, p. 305); the second, and equally important, is that the child "renounces his desire when the spatial connection seems to him insufficient" (Piaget, 1971a , p. 292).

Both these indexes can be more easily verified in contexts where the child has to recover objects that are out of his reach by means of intermediaries. Two such conditions are those encountered in the "support problem" and the "stick problem". Though, as we shall see, they present quite different additional difficulties, both these problems come to be mastered by children during stage five of sensorimotor development.

The behavior pattern of the supports is the first interesting case: drawing some bulky object (cushion, coverlet, etc.) toward oneself to reach objects placed upon it. As we have seen, this behavior is at first purely phenomenalistic. In grasping the support the child sees the object move. Thus he establishes a link of cause and effect between the movements of the support and those of the object. But at its point of departure this link is spatialized so little that the child even draws the support toward him when the desired object is placed beside it. Up to then causality remains characteristic of the fourth stage—intermediate between phenomenalistic efficacy and truly spatialized causality. On the contrary, to the extent that the behavior pattern of the support has become systematic it is spatialized, and form the beginning of the second year of life gives rise to connections typical of the fifth stage" (Piaget, 1971a, p. 319).

In the support problem, therefore, the relation of spatial contact between the target and the intermediary object is established by the experimenter (while in the stick problem, it has to be actively established by the subject): what we want to see is whether the subject recognizes it as a necessary condition for the existence of a causal effect of the intermediary on the target object. If this is the case, then the subject should pull the support when the target object is placed on it, but should, at the same time, resist pulling it when the target object is clearly separated from it.

In the few instances in which this problem has been offered to nonhuman primates positive results have been reported. Redshaw (1978) asserts that gorilla infants begin to be capable of pulling a support to obtain a reward at about 6 months of age, but only two months later they resist pulling it when the reward is held suspended 4 in above it. Mathieu, Daudelin, Dagenais and Decarie (1980) only state that their two chimpanzees solved the support problem at 28 an 30 months, and that one of them, who was also tested at 22 months, did not solve it.

Both these studies do not give methodological details on type and number of trials given and both seem to focus more on the positive achievement of pulling the support to obtain the reward than in probing the more important understanding of the necessity of spatial contact. Furthermore, as in all other cases where evidence for a capacity comes from problem solving, one should always try to control, as far as possible, that the solution is not being obtained by possible alternative strategies. For example, if the relative positions of the support and the target object are fixed and/or very few in number and trials are repeated many times, the subject might easily learn and memorize in which global situation his action is successful and in which is not, and behave consequently. The problem would be correctly solved but nothing could be said about the subject's understanding of the spatial contact condition. As we shall see our testing situation tried to incorporate several controls again such "blind" strategies.

METHOD

Subjects

Subjects were the Japanese macaque Oi at 15 months, the gorilla Rm at 19 months, the cebus Ro and Pe at 9 months. These are the same subjects of the longitudinal studies reported in chapters 3–5. Two of the crab-eating macaques that took part in the study of object-concept (chapters 7 and 8), Pi at 31 months and Pa at 33 months, were also tested.

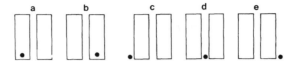

FIG. 1. Diagram of the five conditions presented in the first support test. The black dot indicates the position of the reward.

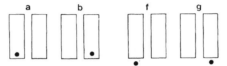

FIG. 2. Diagram of the four conditions presented in the second support test. The black dot indicates the position of the reward.

Procedure

All subjects were familiarized with the test situation by offering them for a few minutes a day for a week the following condition. On a wood board (50 × 70 cm) attached in front of their cage bars a rectangular strip of cloth, measuring 10 × 30 cm for the gorilla and 5 × 30 cm for all other subjects, was presented, perpendicularly to the cage bars, either with a small candy placed on it at its distal extremity (with respect to the subject's position) or without. The candy was out of the direct reach of the subjects and could be secured only by drawing the support. The first condition was offered to familiarize the subject with the possibility of obtaining the reward by means of the support, and the second to allow the manipulation and exploration of the strip itself so that they would be less likely to draw it just for the sake of curiosity toward a novel object. Notice that the crucial condition in which the reward is not placed on the support was not given in this phase.

After this familiarization phase Test 1 was administered, individually to each subject. Two strips of cloth of the same dimensions as those used in the pre-testing phase were placed, parallel to each other, on the wood board at a distance of 10 cm from each other. A reward, consisting of a small candy, was placed in one of the following five positions: (a) and (b), on one of the two strips at its distal extremity; (c) and (e), on one of the two external sides of the two strips, at its distal extremity and at a distance of about 3 cm from it; (d), between the two strips, at their distal extremity (the five positions are diagrammed in Fig. 1). In all five positions the reward was on the same ideal horizontal line.

The first two subjects, Oi and Rm, received 20 trials (four for each of the five positions) in each session. 4 sessions were administered to Rm and 3 to Oi. All other subjects received, instead, 24 trials per session (four for each of positions c, d, e, and six for each of positions a and b), because of a later methodological suggestion to balance the number of cases in which the reward was and was not recoverable. Each subject received three sessions. Order of trial presentation was randomized within each session.

The reason for having two supports rather than one was to be able to have several perceptually different support and reward positions in order to avoid presenting the same position many times and make it more difficult for the subjects to learn to respond to few fixed absolute patterns.

Yet, even with 5 positions, each repeated only 4 times, one cannot exclude completely, in the event of a correct performance, that a simple discrimination learning has taken place. To control for this possibility, a second test was administered to each subject right after the completion of the first. In this test, which was administered in one session only, the most crucial conditions, the ones where the reward was placed outside of the support, were modified in order to create a situation perceptually very different from the one presented in the first test. Rather than on the lateral sides of the support, the reward was placed in front of their distal extremity, as diagrammed in Fig. 2. 24 trials, equally divided among the four positions were given to all subjects, except the gorilla, Rm, which received 18 trials, 9 with the reward placed on the support and 9 outside of it.

RESULTS

Table 1 presents results obtained by each subject in the first test. Obviously, the correct response pattern is to pull the support when the reward is placed on it (conditions a and b) and resist pulling it when the reward is placed outside of it (conditions c, d, and e). The performance of each subject was significantly correct (Simple chi square, df $= 1$, $p < .001$).

They practically never failed to pull the support when the reward was placed on it, but they also resisted pulling it in the outside conditions in the large majority of cases.

Results of the second test are given in Table 2. In this test too, all subjects' performances were significantly correct (Fisher exact probability test). All subjects quickly adapted to the new situation and correctly discriminated the new positions where the reward was placed outside of the support.

DISCUSSION

Beside quantitative results, qualitative observation of the subjects' action also revealed that they were discriminating between the outside and inside

TABLE 1.

Type and Number of Responses to the Inside (a, b) and Outside (c, d, e) Conditions of the First Support Test by Each Subject.

Subject	Inside (a + b)		Outside (c + d + e)		Total
	Pull	No pull	Pull	No pull	
Oi15*	24	-	4	32	60
Pa31*	36	-	4	32	72
Pi33*	36	-	8	28	72
Rm19*	32	-	13	35	80
Ro9*	36	-	9	27	72
Pe9*	34	2	6	30	72

* p < .001

TABLE 2.

Type and Number of Responses to the Inside (a, b) and Outside (f, g) Conditions of the Second Support Test by Each Subject.

Subject	Inside (a + b)		Outside (f + g)		Total
	Pull	No pull	Pull	No pull	
Oi15**	12	-	1	11	24
Pa31**	12	-	4	8	24
Pi33**	12	-	3	9	24
Rm19*	9	-	3	6	18
Ro9**	12	-	1	11	24
Pe9**	12	-	2	10	24

* p < .005
** p < .001

conditions. When the reward was placed on the support the action of drawing tended to be slow and accompanied by careful visual monitoring of the displacements of the reward. On the contrary, when they drew the support in the outside positions, the action was often performed in one rough pull, suggesting a possible reaction to frustration.

The fact that the correct discrimination of the reward positions was immediately generalized to the different situations of the second test makes it difficult to think that subjects had simply learned to respond appropriately to the five specific positions of the first test and, by the same token, makes it reasonable to infer that subjects were indeed guided by the condition of spatial contact between the reward and the support in regulating their response.

We can therefore conclude that all four species of nonhuman primates reach this elementary level of understanding of the operation of physical causality: they can at least recognize when causal relation between two objects is possible on the basis of there being spatial contact between the two objects.

10 Causality II: The Stick Problem

Francesco Natale
Istituto di Psicologia, C.N.R.
Rome, Italy

INTRODUCTION

In the preceding chapter we have investigated the understanding of some of the basic conditions underlying the operation of physical causality through the support problem.

The use of a stick to rake in out-of-reach objects is a task that probes further this understanding, as well as that of the elementary physical relations governing interactions among objects.

Contrary to the support, the stick is a real detached intermediary: it is an independently manipulable object whose relation to the target object needs to be actively constructed by the subject. Consequently, its correct use poses two additional problems. First of all, the condition of physical contact between the intermediary and the target object is not given in the situation, as in the support case, where the subject's task is to recognize it, but must be established by the subject's action. Second, once contact has been established, only a specific action of the subject will enable the recovery of the object. Point of contact, angle of impact and strength of impact are the variables that need to be mastered, in relation to one another, in order for such an action to be successful. Furthermore, the reciprocal relations of these variables are regulated by the principles of the (inverse) lever: the stick acts as an inverse lever amplifying the movement of the distal extremity. This amplification needs to be taken into account in "calculating" the effects of the striking action. These constraints do not exist in the support case, where physical relations between the intermediary and the target object are also given in the situation and only the strength of the pulling action needs to be grossly adjusted.

Hence, the stick problem allows the study of several aspects of basic physical cognition. It allows, furthermore, a direct comparison of nonhuman primates with human children, whose behaviors in this condition have been analyzed by Piaget in his classical longitudinal studies of sensorimotor intelligence. Piaget found that the use of the stick is learned between 13 and 16 months, in stage 5, and described some of the typical behaviors preceding its correct use:

Observation 157. At 1;0 (5) Lucienne already possesses the "behavior pattern of the support", . . . I try to determine, the same day, if she is capable of that of the stick. One will see that she is not. The child is playing with a very elongated cover which can fulfill a stick's function; with it she hits the tiers of her table, the arms of her chair, etc. Then I place before her, out of reach, a small green bottle for which she immmediately has a strong desire. She tries to grasp it with outstretched hands, struggles vigorously, wails, but does not have the idea of using the cover as a stick. I then place the cover between her and the bottle: same lack of comprehension. Then I place the bottle at the end of the cover: Lucienne pulls the cover to her and grasps the bottle as we have observed in Observation 150 repeated. Then I again put the bottle out of reach, but this time I place the cover next to the object and at the child's disposal; nevertheless it does not occur to Lucienne to use it as stick. (Piaget, 1974, pp. 297–298)

Observation 154. B. At 1;1 (0) . . . Jacqueline holds a celluloid doll trimmed with a rattle which I sounds at the slightest movement. I take it from her hands and hide it behind the hedge of the bassinet, at a different place from where it was two hours earlier; Jacqueline then tries to see the vanished doll, leans over and looks for a moment, then, as if an idea occurred to her, picks up the stick lying at her feet and strikes the edge of the bassinet with it at the exact place where the doll disappeared. After a few minutes I return the doll to her and repeat the experiment at several other places. Each time, Jacqueline takes the stick to tap on the edge of the bassinet at the point where the desired object disappeared. It is difficult not to see in such movements a procedure to make the doll return, the causality inherent in the stick therefore, regresses, in such a case, toward pure efficacy (Piaget, 1971a, pp. 320–321).

Observation 161.— . . . At 1;1 (28) [Jacqueline] is seated on the floor and tries to reach the same cat, this time placed on the floor. She touches it with her stick but without trying to make the cat slide to her, as though the act of touching it sufficed to draw it to her. (Piaget, 1974, p. 300)

Observation 161.— Finally, at 1;3 (12) [Jacqueline] discovers the possibility of making objects slide on the floor by means of the stick and so drawing them to her; in order to catch a doll lying on the ground out of reach, she begins by striking it with the stick, then, noticing its slight displacement, she pushes it until she is able to attain it with her right hand. (Piaget,1974, p. 301)

Observation 158.— . . . At 1(0)4 . . . . I place the flask on the floor, 50 cm. away from Lucienne. She begins by wanting to grasp it directly, then she takes the stick and hits it. The flask moves a little. Then Lucienne, most attentively, pushes it from left to right, by means of the stick. The flask is thus brought nearer. Lucienne again tries to grasp it directly, then takes the stick again, pushes it once more, this time from right to left, always bringing the object towards her. Delighted, she grasps it, and succeeds in all the subsequent attempts. (Piaget, 1974, p. 298).

The first of these observations shows that although the child knows the support as an intermediary, and recognizes it when given in the situation, she cannot use even the very same object to actively establish herself its role as intermediary, and consequently she continues to try to reach the target directly. In the second observation, the stick is somehow understood as a possible intermediary, and in fact it is taken and manipulated in relation to the goal of obtaining the target object, but there is no understanding of the contact relation necessary for its operation. Causality has the form called by Piaget "efficacy" or "magico-phenomenalistic": causal links are established directly between the subject's gestures (in this case, the action of beating the stick) and the desired effects, without necessarily requiring physical contact between the cause and the effect. The third observation shows, on the contrary, that contact is understood as an essential condition, but there is no further understanding of the dynamic relations regulating the modalities of this contact.

Finally, the last two observations show how the child gradually comes to master also these relations.

In this study we will follow the development of the use of the stick in macaque, gorilla and cebus.

Previous Studies

Though the use of a stick as a raking tool is one of the most recurrent topic of the literature on tool use in nonhuman primates, results reported are somewhat contradictory. Apes are generally considered proficient in such task and gorillas, in particular, have been reported to use sticks, at least occasionally, by Parker (1968; 1969), Redshaw (1975), Chevalier-Skolnikoff (1977), Beck (1980). Yerkes (1927a), however, in his accurate study of gorilla's cognitive abilities, failed to elicit any consistent use of the stick in his 5-year-old captive subject. Though one year later his subject's performance appeared to have improved, Yerkes (1927b) stated that the use of the stick remained quite primitive and sporadic. The literature on cebus appears more consistent in reporting success in experimental settings (Kluver, 1933; Harlow, 1951; Cooper & Harlow, 1961; Warden, Koch, & Fjeld, 1940). In some of these studies, cebus also used a short stick to get a second stick, long enough to reach the target.

Strongly controversial is instead the literature on macaques. Hobhouse (1926), Shepherd (1910) and Beck (1976) report positive results; Dresher and Trendelenburg (1927), Nellman and Trendelenburg (1926), Watson (1908), Yerkes (1916) report complete failure in using stick or similar objects as raking tools. Somewhere in the middle are Warden, Koch, and Fjeld (1940) and Shurcliff, Brown, and Stollnitz (1971), who found that some shaping was needed to learn to use an object as a reaching tool. Kluver (1933) showed that crab eating macaques were successful in some problems with ropes, while the use of detached objects as tools was almost completely absent.

Even putting aside episodic reports, there are two main problems with this literature. One is the emphasis on performance rather than on understanding. Results are focused on whether and to what extent the animal actually succeeds in obtaining the reward, rather than on analyzing the details of its performance as an indication of how the principles and the variables regulating the use of the stick are understood. As we shall see, and as it nearly always is the case in problem solving tasks, success might or might not be correlated to such an understanding. In the second place, very few of the quoted studies are developmental and hence lack a more general picture of the capacities underlying specific behaviors observed.

Using quantitative criteria of success to define presence or absence of a given ability might satisfy our need for rigor and objectiveness, but it is a far cry from its understanding. We will consequently concentrate more on a qualitative assessment of the subjects' behaviors, of their modifications in the course of development, and, especially, of what they reveal about the subjects' cognition of the physical domain, rather than on the percentages of their successes and failures.

METHOD

Subjects

Three of the subjects tested were followed longitudinally. All of them belonged to the group of subjects whose sensorimotor development had been followed since birth and described in chapters 3, 4 and 5. They were the Japanese macaque (*Macaca fuscata*) Oi, the gorilla (*Gorilla gorilla gorilla*) Rm, and the cebus (*Cebus apella*) Ro. Their behaviors with the stick were observed between 15 and 38 months of age for the macaque and gorilla and between 5 and 20 months of age for the cebus. In a first period (15–18 months for macaque and gorilla, 5–8 months for cebus), the animals were given sticks together with other objects, both from outside and inside their cages, that they could freely manipulate for familiarization purposes. After this period, formal testing began, according to the procedure specified below. The macaque between 19 and 22 months, the gorilla between 19 and 21 months and the cebus between 9 and 12 months, each underwent 12 testing sessions. Following this period, each subject received one testing session per month. Experimental sessions did not have a fixed number of trials; they were paced to the animal's level of interest and cooperation. They lasted each between 20 and 40 minutes and included a number of trials varying from 20 to 32. Subjects' behaviors were manually scored by two experimenters using a style sheet. Some of the sessions were also videotaped to allow a more careful analysis of the behaviors shown.

Five more subjects, two macaques and three cebus, were tested transversally as controls. They were two of the crab-eating macaques (*Macaca fascicularis*) already tested for stage 5 and stage 6 object-concept and in the support problem, Pa and Pi, at the age of 34 months, cebus Ca, of the longitudinal group, at 45 months, and two adult females, born in captivity and belonging to the colony of cebus housed in our lab, Bh at age 66 months and Pp at age 78 months. The macaques received each 7 sessions and the cebus 8 over a three-week period. Some of the cebus data are also reported and discussed in Parker and Poti' (in press).

Apparatus

The experimental apparatus consisted of a 70 × 90 cm wooden platform that could be moved in front of the bars of the subjects' cage at the level of the cage floor. Wooden sticks used as tools were about 30 cm long for the gorilla and 20 cm long for other species. The stick was placed on the platform, perpendicular to the cage bars; the reward, a piece of fruit or a candy, was placed out of the animal's reach, either on the right or on the left side of the distal extremity of the stick, at about 5 cm of distance.

Two extra positions were added in testing cebus: the reward was placed in front of the distal end of the stick at about 5 cm of distance (condition "c" in Figure 1); the stick was placed parallel to the cage bars between the subject and the reward (condition "d" in Figure 1).

All positions used are diagrammed in Figure 1. After positioning stick and reward on the platform, the apparatus was moved to the front of the cage within the subject's reach. The subject was allowed to make a response

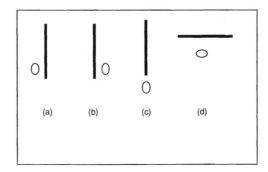

FIG. 1 Diagram of the four relative positions of stick and reward used in the stick task. Subject is placed at the position indicated by the top line.

and then the apparatus was withdrawn and reset for a new trial.

RESULTS

The behaviors shown by our subjects can be divided into three groups, that closely match those seen in children, but with significant differences among species and in developmental phases.

In the first group we class all behaviors where no systematic attempt of contacting the reward with the stick was made. All subjects started by trying to reach directly the reward stretching their arm through the cage bars. The stick was either ignored or taken and manipulated for its own sake. These included biting and chewing on the stick, waving it in air, or simply displacing it on the platform without grasping it. No connection was established between the stick and the reward. After the first sessions, however, some of the actions performed with the stick began to show some orientation toward the reward.

The macaque grasped the stick at one end and slid it on the platform parallel to its initial position or horizontally so to sweep the board in a single movement. However, it did so both toward the reward and away from it, without any rule or sistematicity, possibly depending on the hand with which it happened to grab the stick. This action resulted sometime in hitting and displacing the reward, and though most of the times the reward was projected outside of the platform, it ended sometime within the reach of the animal that recovered it. Between 19 and 22 months this behavior did not undergo any modification, and, especially, it continued to be directed randomly both toward and away from the reward.

The gorilla between 19 and 23 months showed a similar behavior; it also threw the stick in the direction of the reward and it banged it on the platform. Some of the actions accidentally resulted in bringing within reach the reward, but this did not result into any improvement of the actions themselves.

Between 5 and 8 months the cebus performed two main actions with the stick: it drew it carefully inside the cage while looking at the reward and it hit it on the platform while looking at the reward.

Between 9 and 14 months cebus' actions with the stick became more oriented toward the reward. It slid the stick over the platform and it made rotate its distal extremity by pivoting on its proximal extremity. However, these movements were performed both toward and away from the reward. They were not systematically aimed at establishing contact between the stick and the reward. Magic-causation behaviors, such as hitting the stick on the platform or rubbing it behind the reward, continued to be performed. Cebus' action was, however, in general less violent and more accurate, and, especially toward the end of this period, more and more successful in contacting the reward. For example, it developed a strong right-hand preference that led it to rotate the stick almost always counterclockwise; however, when the reward was on the right side of the stick and would have been missed by its counterclockwise rotation, the animal attempted sometime to reposition the stick on the right of the reward.

After this first period, the behavior of all subjects underwent substantial changes, but with different results in the three species. Gorilla (at 24 months) and cebus (at 15 months) began to establish contact systematically between the stick and the reward, whatever the action involved. This systematicity is confirmed by the disappearance of all the actions where the stick was moved away from the reward, and, in cebus, also by the disappearance of all the magic-causation behaviors. The stick is now always moved toward the reward and, except for some occasional miss, it nearly always strikes it.

The macaque, on the other hand, developed, from the age of 23 months, a stereotyped action pattern that it applied almost systematically: it made rotate the distal extremity of the stick by pivoting on the proximal extremity with its hand. However, it did so almost always with its left hand and from left to right (i.e., impressing a clockwise rotation), irrespectively of whether the reward was placed on the right or left hand side of the stick. This action was often successful in bringing the reward within reach, but only when this happened to be placed on the right side of the stick. It was obviously completely unsuccessful when the reward happened to be placed on the left side of the stick.

This behavior did not improve, and did not undergo any change in pattern, throughout the end of testing at 38 months. Notice that, as a consequence of these changes, all subjects increased their success rate with respect to the first period: the macaque passed from 8% to 33% of successful recovery, the gorilla from 7% to 41% and the cebus from 26% to 58%. They did so, however, for quite different reasons: gorilla and cebus by implementing systematically the condition of contact between the stick and the reward, the macaque by applying a "blind" stereotyped strategy. In fact, if we divide the macaque's successes according to whether the reward was placed on the right or left hand side of the stick a large asymmetry emerges: when the reward was placed on the right hand side of the stick successful recovery took place 37% of the times, when it was placed on the left only 6%. No such difference appeared in the gorilla's and cebus' performance.

Though the gorilla correctly hit the reward with the stick nearly always, his action was very often unsuccessful in bringing it within reach because of persistent difficulties in regulating the angle of impact between the stick and the reward and the force impressed to the stick. More specifically the angle was too narrow (i.e., the stick was too close to its original perpendicular position) and the force impressed was far in excess. As a result, the reward was displaced too violently and/or with a too horizontal trajectory, ending out of reach or outside of the platform after the first or second hit. Both these "errors" seem to be related to a lack of understanding that the movements impressed by the hand at the near extremity of the stick are amplified (proportionally to the length of the stick) at its distal extremity. Thus, if the stick is perceived as an extension of the hand and its movements are regulated as if it were the hand to hit the reward, without taking into account this amplification effect, the resulting effect will be overscaled. To obtain a correct effect hand movements must be decreased in strength and the angle of impact must be increased proportionally to the length of the stick, or, more precisely, to the distance between the hand and the point of impact. In other words, a covariation relationship among these three physical variables must be constructed. This did not happen to the gorilla, whose actions continued to show this overscale effect till the end of her third year of life. At this time an entirely new behavior pattern appeared. She took her "Linus-blanket", that she used to carry around in her cage, and threw one of its extremities through the cage bars onto the platform toward the reward, so to cover it. Then she pulled the cover back and in so doing made the reward slide toward the cage and recovered it. From that moment on she threw the blanket more and more often, with progressive accommodation of the scheme: she drew the blanket back more and more carefully in order not to slip over the reward. She did not stop entirely to attempt with the stick, but would typically resort to the blanket after one or two unsuccessful trials with the stick. These behaviors stayed unchanged till the end of the trials at 38 months.

Between 15 and 17 months, the cebus showed much the same difficulties as the gorilla in maneuvering the stick. Contact with the reward was nearly always established, but the action was often unsuccessful because of the same errors in calculating the point, angle and effort of the impact. Errors were in the same overscaled direction, indicating the same difficulties in mastering the physical relations among the objects. Cebus performance, however, definitely changed and improved over time.

Beginning at 18 months the physical variables involved in the problem appeared to be substantially mastered. Hence the action became appropriate in drawing most of the times the reward within reach. This is also confirmed by a change in rate of success in recovering the reward, that increased from 58% between 15 and 17 months to 88% between 18 and 20 months.

These results were largely confirmed by the transversal studies. Neither of the two macaques arrived to establish systematic contact between the stick and the reward; i.e., they only showed the first group of behaviors. Over the course of trials, Pa developed the same stereotyped strategy seen in Oi, only on the opposite side: it would rotate the stick counterclockwise with its right hand, independently of whether the reward was placed on the right or left hand side of the stick. Consequently, its pattern of successful recoveries is highly similar to that of Oi: in the second half of the sessions (last 3 sessions) its overall success rate was 33%. It was, however, 60% when the reward was placed on the left side of the stick and only 7% when it was placed on the right. Pi, on the other hand, never developed any strategy and it was constantly unsuccessful, except for some accidental displacements of the reward that happened to end within reach: its overall success rate over the last 3 sessions was 5%.

Strikingly different was the performance of cebus Ca. From the very first session, after a few trials of adjustment, it was able to rake in the reward correctly without any difficulty. The reward was systematically contacted since the beginning, and the physical variables were quickly adjusted to the right proportions. Its overall success rate in the last 3 trials was 86%.

Cebus Pp showed a performance pattern similar to that of Ca. Only it took it a longer time (the first two sessions) in order to appropriately adjust the action of the stick. Its overall success rate in the last 3 sessions was 87%. Cebus Bh, on the other hand, appeared in general to be scarcely motivated to work with the stick. During the first two sessions, it performed a number of irrelevant manipulations with the stick and then slowly developed a stereotyped strategy similar to the one seen in macaques Oi and Pa. When the reward was placed either on the left or on the right-hand side of the stick, it began to rotate the stick systematically clockwise, thus succeeding in recovering it when it was placed on the right and failing when it was placed on the left. In fact, its pattern of successes resulted similar to that of the two macaques. Over the last 3 sessions, it reached an overall measure of 31%; it was, however, 83% when the reward was placed on the right and 5% when it was placed on the left. The stereotyped clockwise rotation of the stick was, however, not as systematic as in macaques. When the reward was placed in front of the stick, a number of other actions, like sliding or moving back and forth or pulling, were performed. None of them was, however, appropriate to the correct displacement of the reward.

DISCUSSION

The behaviors shown in the stick problem reveal large differences both among the nonhuman primate species tested and with respect to the human infant. Though types of behaviors shown are similar to those seen in children and denote levels of understanding of the causality mechanisms also similar to those of children, their ontogenetic evolution is quite different and it is different in each species.

When the stick begins to be perceived as a possible intermediary to reach the reward, all subjects show the typical magico-phenomenalistic understanding of causal relations. The subject's action on the stick is considered efficacious per se, with no understanding of the physical conditions required by a cause-and-effect connection, not even the most elementary one, that of spatial contiguity. Consequently, there is no systematic discrimination between actions that might result in contacting the reward with the stick and actions that cannot have such a result. This phase lasted up to 22 months for the macaque, 23 for the gorilla and 14 for the cebus. It is interesting to notice that the same three subjects solved the support problem, which also requires the understanding of the spatial contiguity condition, at 19 (macaque), 19 (gorilla), and 9 (cebus) months. This lag (or "decalage", in Piaget's terms), as reported above directly in Piaget's quotations, is also found in children, and reflects the distinction between passively recognizing when the condition is satisfied and actively establishing it by manipulating an independent object.

Yet this lag appears to be quite large in our subjects: 4 months in the gorilla and 5 in cebus (the macaque never arrived at systematically establishing the contact condition). Furthermore, it should be added that, as we saw in chap. 7, the macaque and gorilla had reached stage 5 of object concept development at 18 months and the cebus at 7 months.

Even larger differences characterize the second phase of this development. The passage from a magico-phenomenalistic to an objective and spatialized conception of causality never took place in the macaque. The understanding that the action of the stick can cause the reward to be displaced only if the stick comes at least in contact with the reward was never realized. As in other tasks (for example, in the invisible displace-

ments, chap. 8), the persistent incomprehension of the conditions governing the appropriate action led this animal to develop an ad hoc, practical
strategy that was randomly successful in obtaining the reward. Henceforth,
it "economically" adhered to it and stopped to attempt any another action
pattern. One of the two older macaques tested followed an almost identical
course, while the other did not even develop a fixed, task-specific strategy,
at least in the course of a comparable number of trials. Gorilla and cebus,
on the other hand, overcame the magico-phenomenalistic phase and reached
a fully spatialized conception of causality. It should be noticed that this
fundamental cognitive achievement is not revealed by their increased success rate in obtaining the reward, which, at least for the gorilla, is not
different from that of the macaque, but by the disappearance of those
actions with the stick that are contradictory with the understanding of the
necessity of the contact condition (like displacing the stick in the direction
opposite to the reward, or beating it on the platform).

Once the human infant realizes that the stick must necessarily strike the
target object to displace it, it immediately proceeds to adjust appropriately
the physical variables governing the resulting trajectory of the target. This
process might take from a few practicing trials to a few weeks (for example,
two and six weeks, respectively, in Piaget's own subjects, Jacqueline and
Lucienne, as reported in the Observations quoted above). Our gorilla, on
the other hand, never succeeded in doing so: practicing for the whole third
year of life did not lead it to master the mechanical constraints governing
the operation of the stick. A wonderful demonstration that its difficulties
resided exactly in mastering these constraints was offered by its spontaneous substitution, after 15 months of attempts, of the stick with the blanket. Contrary to the stick, using the blanket as a raking tool does not
require any of the complex mutual adjustments of point, angle and force
of impact of the tool with the target: once contact between tool and target
has been established (in the simple form of covering the comparatively
much smaller target with, practically, whatever portion of the blanket),
the only action that needs to be grossly adjusted is that of drawing the
blanket, which, in turn, reduces to simply avoiding a single violent draw
that would make the blanket slide over the target. In other words, the
blanket offered to our subject the possibility of circumventing just those
mechanical and dynamical constraints that it was not capable of mastering.
Quite surprisingly, only the cebus did complete this second phase and
became capable of mastering the physical variables involved in the use of
the stick. It took it, however, more than three months of practicing efforts:
its whole development in this domain appeared to be quite slow and retarded with respect to the human infant. As it has already been remembered, this subject showed stage 5 abilities in the domain of object-concept
at 7 months and mastered the support problem at 9 months, while it reached

proficiency in the use of the stick only at 18 months. This subject appears not to be an exceptional, isolated case. Both Ca, at more advanced age and the adult Pp were capable of mastering the stick problem after a short practice, which seems to indicate that adequate preconditions for mastering the physical variables involved in the problem had already developed in these subjects. The second adult subject, however, turned out to be incapable of mastering the same problem, even at the level of its most elementary preconditions. If we bar possible questions of motivation (which could only be judged subjectively) or some specific anomaly of this subject (which we have no indication of), this failure might indicate that the level of capacities required by the stick problem, though within the reach of this species, is close to its upper limit.

CONCLUSION

It is hard to resist noticing the perfect correlation existing between these results and those that emerged from the analysis of the whole longitudinal course of sensorimotor intelligence development of the same subjects, as discussed in part II of this book. The developmental pattern noticed there, consisting of a selective lack of development of the physical domain of cognition in comparison to the human infant, beginning as early as stage 3 of sensorimotor development with the absence of secondary circular reactions, and progressively extending to that of secondary coordinations and exploration of object interactions, would predict the impossibility, or extreme difficulty, of constructing the causal and dynamical relations governing the physical interaction of stick and target object. The differences among species seen in this cognitive domain, with the near complete and persistent lack of development of the macaque, the strongly limited development of the gorilla, and the more advanced, albeit strongly retarded, development of the cebus, are perfectly matched by the different levels of proficiency achieved in mastering the physical variables of the stick problem by the same three species.

The gorilla's spontaneous "invention" of the blanket as a raking tool, on the other hand, seems to confirm the result obtained in testing for stage 6 object-concept (see chap. 8): this animal seems to possess some representational capacity. The way in which it used this tool for the first time is completely analogous to the behaviors described by Piaget (1974) as instances of "invention of new means by mental combination", and considered by him as a typical mark of stage 6 representational abilities. This animal, in fact, suddenly dropped the stick, abandoned the platform with the reward and went to look for its blanket in the back of its cage. Then it came back to the front and, without any hesitation, threw the blanket through the cage bars on the reward: the role and use of the blanket were mentally figured out before any action and when the blanket was not even present. The striking peculiarity of the gorilla, as opposed to the child, is that representation is not used to enable the mental (and hence faster and more powerful) combination of variables of more complex problems, as typical of stage 6 children, but rather to evoke a cognitively simpler and more primitive situation. In other words, the advanced "form" of mental representation is applied to a primitive "content". This peculiarity is again a reflection of the "mosaic" developmental pattern of the nonhuman primate species (obviously, relatively to a human "norm"), whereby a given level in one domain of cognition does not correspond to a comparable level in another domain.

Many years ago, one of the founding fathers of the comparative study of intellectual capacities, W. Kohler, while investigating the way chimpanzees solve problems requiring the use of intermediaries, and, more specifically, the stacking of boxes to reach a target hanging from the ceiling by climbing on them, offered a penetrating account of where exactly seem to lie the greatest difficulties for these animals:

> I want to emphasize again that at first everything goes well: as soon as the animals are quite familiar with the situation and are convinced that they cannot obtain the objective with *one* box, a moment arrives when the second box is suddenly "drawn into the task". They drag it up (Tschego) or carry it just to the first box and all of a sudden stop and hesitate. With uncertain movements they wave the second one to and fro over the first (unless they let it drop to the ground immediately, not knowing what to do with it, as Sultan once did) and if you did not know that the animals see perfectly well in the ordinary sense of the word, you might believe that you were watching extremely weak-sighted creatures, that cannot clearly see where the first box is standing. Especially does Tschego keep lifting the second box over the first and waving it for quite a while, without either touching the other for more than a few seconds. One cannot see this without saying to oneself: "Here are two problems; the one is not really a difficult task for the animals, provided they know the use to which boxes can be put; the other (*add one box to another so that it stays there firmly, making the whole thing higher*) *is extremely difficult* . . . here the chimpanzee meets *a problem of statics*. . . . the total impression of all observations made repeatedly on the animals leads to the conclusion that *there is practically no statics to be noted in the chimpanzee*. . . . While human children, when about three years old, begin to develop the elements of this naive physics of equilibrium, the chimpanzee does not seem to make any essential progress in this direction, even when he has plenty of opportunity to practise" (Kohler, 1976; pp. 147–150).

Seventy years later, it would be hard to find more appropriate words to describe the specific nature of the mixture of competence and incompetence seen in these animals.

IV STRUCTURE AND DEVELOPMENT OF LOGICAL COGNITION

11 Introduction to the Study of Logic

Francesco Antinucci
Istituto di Psicologia, C.N.R.
Rome, Italy

The longitudinal studies of developing sensorimotor intelligence in our nonhuman primate species showed (in comparison to the human primate), besides more obvious quantitative differences in times and rhythms of development, the existence, at a very early stage, of a striking qualitative distinction deeply affecting the process of development: the establishment of an unbalance (beginning as early as the third stage) between the development of the logical domain of cognition and that of the physical domain (see chapters 3–6). This was confirmed by the follow up studies presented in the preceding chapters, through the analysis of development of specific cognitive categories, like the concept of object and that of causality. Development in the physical domain appeared either lacking, beyond anything more than an elementary level, or strongly limited, or, at least, extremely retarded and laborious.

This result appears all the more interesting in view of the fact that the distinction between logical and physical knowledge has a fundamental status within the piagetian theory of cognition. As outlined in the introduction, these two types of knowledge have two different origins in the interaction of the cognizing subject with the surrounding environment and leave their organizational mark in the construction of all later cognitive structures. To stress the point once more:

> Examination of the child behavior in regard to objects shows that there exist two kinds of experiment and two kinds of abstraction, depending on whether the experiment is based on things themselves and allows for discovery of some of their characteristics, or whether it is based on coordinations, which

were not in things but that the action, in utilizing the latter, had introduced for its own requirements. . . . there is the experiment on the object leading to an abstraction from the object. This is the physical experiment which, properly speaking, is a discovery of the characteristics of things. . . . For example, in picking up solids, the child will notice by physical experience the diversity of weight, its relation with volume of equal density, the variety of density, and so for. On the other hand, the child who counts ten pebbles and discovers that they are always ten even when he permutes the order, does an experiment of an entirely different nature. Actually, he experiments not on the pebbles, which he uses merely as instruments, but on his own actions of order and enumeration. In this sense, knowledge is abstracted from action as such and not from the physical characteristics of the object. . . . order has been introduced by action on the pebbles (arranged in a row or in a circle) as well as their sum itself (due to an act of colligation or reunion). What the subject then discovers is not a physical characteristic of pebbles but an independent relation between the two action of reunion and ordination . . . the experience is authentically logico-mathematical. (Piaget, 1972, pp. 29–30; 70–71)

Notice, however, that the status of these two kinds of knowledge in cognitive organization is by no means symmetrical:

If pure logico-mathematical knowledge exists detached from all experience, reciprocally there is no experimental [physical] knowledge that can be qualified as "pure", detached from all logico-mathematical organization. *Experience is accessible only through logico-mathematical limits consisting of classifications, functions, and so forth.* (Piaget, 1972, pp.72–73, italics added).

An example of this dominance was seen in the analysis of the stick problem. The proper mastering of the physical variables involved in correctly hitting the target with the stick requires, beyond the appreciation of the different physical effects produced, the establishment of a covariation relation (direct and inverse) (i.e., an elementary function) between the variables themselves (see chap. 10).

In view of this "primacy" of the logical component in acquiring and structuring knowledge, lack, limitation and retardation of development in the physical domain might have two (and, by no means, mutually exclusive) origins. On the one hand, it might be generated by specific lack or limitation of action oriented to "physical experiences", as defined above, i.e., by limitations of physical cognition *strictu senso*. That there are such specific limitations in the nonhuman primate species we tested has been well evidenced by the analysis of the first four stages of sensorimotor development (lack of secondary circular reactions, lack of object-object interactions, etc.). On the other hand, it might be generated or furthered by an insufficient development of the logical support necessary to organize physical knowledge. This, of course, would be true in the more complex cases of such organization, where more than an elementary logical support is required (and the stick problem might well represent the lower limit of such complexity).

Finally, both these processes might take place. Is, for example, the cebus slowness and retardation in appropriately mastering the stick problem, and its occasional failure, also due to a slower and, possibly, limited evolution of the logical supporting structures?

The longitudinal studies of early developing cognition has little to say in this respect. In fact, only the very initial precursors of logical structures are generated during this period, such as sequential schemata coordinations or the more advanced hierachical coordinations, as seen, for example, in means-end coordinations. Contrary to the physical domain, it is true that these developments did not appear different than in the human child, but we know very little about the further impressive developments that, in the human child, occur after 10–12 months. For all these reasons, it seemed to us that, in order to clarify and understand better our comparative results, further probes were needed, and especially in determining the further development and organizational level achieved by nonhuman primates in the domain of logical cognition. In doing so, we wanted, obviously, to maintain full comparability of the nonhuman primate species to the human primate, as we did all along our studies.

However, by the time he reaches the sixth stage of sensorimotor development, at about 18 months, the human child develops representational capacities and becomes capable of mental actions on mental objects or symbols. Consequently, he will progressively structure his logical action on this level and it is this process that traditional studies of logic development have examined (for example, Piaget, 1964, 1976; Vygotsky, 1962). Obviously, in Piaget's conception this process must be based, both epistemologically and ontogenetically, on (epistemologically and ontogenetically) prior sensorimotor action:

> We therefore come to our final hypothesis, that the origins of classification and seriation are to be found in sensori-motor schemata as a whole (which include perceptual schemata as integral parts). Between the ages of 6–8 and 18–24 months, which is well before the acquisition of language, we find a number of behaviour patterns which are suggestive both of classification and of seriation. . . . The fact that we can observe various prototypes of classification and seriation at the sensori-motor and preverbal stage of development proves that the roots of these structures are independent of language. (Inhelder & Piaget, 1969, pp. 13–14)

These "roots", however, he did not systematically examine. On the other hand, none of our subjects, with the exception of the ape, ever reached a representational level (see chap. 8) in their continued development, and hence a meaningful comparison can only be carried out at the level in which logical structures are organized in action. Furthermore, if we do this, a crucial question for the phylogeny of human cognitive capacities might be answered by the results of our comparison: Does the large later difference between the human and the nonhuman primates originate fundamentally in the presence vs. absence of development of representation, enabling the power of unconstrained mental manipulation of symbolic objects (in which case we would expect substantially parallel developments in the structuring of nonrepresentational logical cognition between human and nonhuman primates)? Or does it originate in the very process of structuring cognition already at the action level and independently of the additional possibilities offered by representation?

This question is analogous to the one posed before in studying early sensorimotor development: Does the difference between human and nonhuman primates consist in the fact that the latter "stops before", i.e. does not proceed beyond a given stage of sensorimotor development, or are the two courses of development already divergent? As we saw above, it is the second alternative that seems to be correct, and, as we shall see below, it is the same alternative that seems to be true for the development of logical cognition too.

Our target is therefore to compare the development and structure of logical cognition at the level of its original construction in action. Though Piaget's own work has never been concerned more than episodically with this topic, in recent years a number of students have been able to investigate successfully the origins and development of logical categories and operations in the action of human infants from six months on (Sinclair et al., 1982; Forman, 1982; Langer, 1980, 1986).

In these studies the infant is typically presented with an array of objects, more or less structured in ways to provoke and probe the nature of its action, and is then left free to interact with them. Appropriate detailed microanalysis of his actions, their temporal features, the products they construct, and their sequential organization shows their type and degree of structuring.

This approach is also methodologically extremely appealing to deal with nonhuman primates. Being completely non-symbolic and non-verbal, it bypasses the difficulty of devising non-verbal equivalents of tasks with a verbal component, a road frequently fraught with arbitrariness and uncertainties. Second, by relying on the spontaneous activity of the subject, it eliminates the ambiguities inherent in the interpretation of results coming out of the typical problem-solving situations, within which these problems have been traditionally investigated. In experimental paradigms employing tasks such as matching-to-sample, discrimination, oddity, etc. results are usually obtained after long training sessions where the monkeys' responses are patiently shaped to the desired set and then crucially discriminating trial cases are presented. Evidence coming from such experiments is always indirect and open to question, since the possibility of cueing or irrelevant solving strategies can never be completely ruled out. Third, if the origin of logical knowledge is to be found in the coordinations of the subject's manipulative actions, then this approach offers the possibility of provoking and observing directly the very behaviors that generate such knowledge. Finally, since monkeys easily engage in spontaneous object-manipulation results thus obtained can be directly compared to those of children.

Especially noteworthy in this context is the work of Langer (1980, 1986), whose investigation is on the one side exhaustive of all the major logical structures, whose development is detailedly traced from six months on, and on the other side fine grained enough to turn it into a useful scale of comparison. For these reasons, our study is directly inspired by his work and has been kept as close as possible to it, both in the method used and in the analyses performed. These have been divided in three parts.

The first chapter (12) will report the analysis of manipulations of objects as such: what types of manipulatory schemata are applied to objects, their effects and how they are distributed over different kinds of objects. In this way, information will be provided on logical classification at its most elementary level: how and to what extent types of objects are assimilated by different action schemata depending on their properties. Furthermore, manipulations producing causal and dynamical effects on objects and object-interactions will provide information on physical cognition: their analysis will further and offer a more systematic basis to that of the early longitudinal data. Subsequent chapters will examine the products of such manipulatory activity and the relations action establishes among such products. Chapt. 13 will examine whether and to what extent sets of objects created by the subjects' action embed classificatory structures. Chap. 14, finally, will examine what types of logical relations are constructed among sets.

All data for the above analyses were derived from the same free manipulation sessions given to our subjects. Each of them was allowed to play freely with logically structured sets of six objects at a time while a color video camera recorded all their interactions with the objects and the effects produced for a standard interval of time. No previous training, no provoking or stimulating, no rewarding or deprivation was used before, during or after testing sessions: monkeys retained their standard living routines. Subjects tested, materials used and procedure followed are detailed below, while each chapter will separately report the specific way in which data thus obtained were analyzed.

METHOD

Subjects

Unfortunately, our only ape subject, the gorilla female, who did not belong to our lab but to the Rome Zoo, was not available for this study. The study had to be confined to the macaques and cebus housed in our lab. These were: the two crab-eating macaques (*Macaca fascicularis*) Pi and Pa, also tested in the experiments on object-concept, and on the support and stick problems; cebus Ca and Ro, followed since birth and also tested in other experiments; cebus Br, born in captivity, belonging to the colony of cebus housed in our laboratory. Because of the different ages of our subjects, and in order to generate coherent developmental data, the testing design was partly longitudinal and partly cross-sectional. The two macaques were both tested at 22 months the first time and a year later, at 34 months, the second time. Cebus Ca was tested three times, at 16 months, at 36 months and at 48 months. Cebus Br was tested two times, at 36 months and at 48 months. Cebus Ro was tested at 16 months. In this way, data from two age points (22 and 34 months), each with two subjects, were obtained for the macaques, and data from three age points (16, 36 and 48 months), each with two subjects were obtained for the cebus.

Design

Objects presented for manipulation belonged to a set of 16 objects making up a 4 × 4 classificatory matrix: there were four different forms (cups, rings, crosses and sticks) each realized in four different materials (wood, plexiglas, pvc and wood covered with plastic quartz). They were small enough to be easily manipulated with one hand and combined with each other: maximum dimensions were 5 × 5 cm.

In each trial a set of 6 objects was presented, which belonged to one of the following three class conditions.

In the Additive condition the set was composed of two classes of three identical objects each, that differed in only one of their two properties, either form or material. For example, 3 wood cups and 3 wood sticks (difference in form), or 3 wood cups and 3 plexiglas cups (difference in material).

In the Disjoint condition the set was composed of two classes of three identical objects each, that differed in both form and material. For example, 3 wood cups and 3 plexiglas rings. In the Multiplicative condition the six objects presented partially overlapped in both form and material, making up a 3 × 2 matrix. They belonged either to 3 forms and 2 materials (for example, a pvc and a wood cross, a pvc and a wood ring, a pvc and a wood stick), or to 3 materials and 2 forms (for example, a pvc, a wood and a plexiglas ring and a pvc, a wood and a plexiglas stick). In this condition, therefore, no two objects were identical.

Each subject received a total of 24 trials, 8 for each class condition. In the Additive condition, 4 trials were different in form and 4 trials were different in material. In the Multiplicative condition, the matrix was 3 forms by 2 materials in 4 trials, and 3 materials by 2 forms in 4 trials. Within each condition forms and materials presented were balanced across trials, so that each of the 16 different objects was presented an equal number of times.

The test was administered in 8 sessions, one per day, of three trials each, one for each class condition. Each subject was tested alone in an indoor room adjacent to its living cage. In each trial the six objects were scattered randomly on the floor before releasing the subject into the room. The subject was then allowed to manipulate freely the objects for a 5 minute period, beginning from its first contact with them. All trials were recorded on a color VTR placed and operated from outside the experimental room.

Transcription

A complete and detailed transcription of all the recordings was effected noting: (a) the type of manipulation performed and the organ with which it was performed; (b) the temporal sequence of manipulations, their durations and their intervals; (c) the identity of the object(s) manipulated or in any way contacted both directly and with other objects; (d) the spatial position of the objects in relation to one another both before and after each manipulation.

All transcriptions were made by two observers and required typically several normal and slow-motion passages of the sequences being transcribed. Cases of disagreement were resolved through patient frame-by-frame analysis.

Data for each specific analysis were extracted from this master transcription and coded in ways described in each of the following chapters.

12 Patterns of Object Manipulation

Francesco Natale
Istituto di Psicologia, CNR
Rome, Italy

INTRODUCTION

Before analyzing the products of manipulatory activity from the point of view of cognitive organization (that is, the extent and way in which logical relations are mapped onto objects and collections of objects and their transformations) we shall take into account the manipulatory activity in itself. Two fundamental kinds of information can be derived from the analysis of manipulatory acts.

On the one hand, they constitute the epistemological and ontogenetic basis for equating and differentiating objects, and, hence, the basis of logical classification.

> A child [between the ages of 6–8 and 18–24 months] may be given a familiar object: immediately he recognizes its possible uses; the object is assimilated to the habitual schemata of rocking, shaking, striking, throwing to the ground, etc. If the object is completely new to him, he may apply a number of familiar schemata in succession, as if he is trying to understand the nature of the strange object by determining whether it is for rocking, or for rattling, or rubbing, etc. We have here a sort of practical classification, somewhat reminiscent of the later definition by use. But this rudimentary classification is realized only in the course of successive trials and does not give rise to a number of simultaneous collections. (Inhelder & Piaget, 1969, p. 23)

By exhaustively comparing the total universe of manipulations applied to each of a precisely specified and controlled set of identical, similar and different objects, we will try to give a more precise and quantitative account of how classes of object are formed and differentiated from one another through the differential application of schemata to them.

On the other hand, the very nature of the schemata applied to objects and the consequent effects they produce on them can give us an idea of what kind of properties are being explored and, hence, what kind of knowledge is being constructed through their transformative action. In this sense, the analysis of object manipulations will be nothing more than the continuation and, hopefully, a verification on a more precise, controlled and quantitative basis, of what we have been doing in the longitudinal studies, where we followed the progressive construction of cognitive domains through the more qualitative assessment of crucial landmarks in the development of sensorimotor schemata. In fact, the results obtained there and in the subsequent studies will guide us in paying particular attention, in the analysis, to those dimensions that have already strongly emerged as differentially characterizing the course of developments of the different species.

To this purpose, it is essential to establish a theoretically appropriate framework for interpreting the (possibly large) variety encountered when considering the full range of schemata displayed in manipulating objects. Since, as we saw above, a crucial differentiation seemed to emerge in the construction of the physical domain of knowledge, we shall try, as much as empirically possible, to identify categories of schemata according to the type of contribution they predominantly (but, obviously, not exclusively) bring to this domain of knowledge. In this regard, a first elementary taxonomy of schemata was established by Piaget himself in distinguishing among different types of "circular reactions":

> The "secondary circular reactions" prolong, in effect, without adding anything to, the circular reactions under examination hitherto [i.e., "primary circular reactions"]; that is to say, they essentially tend toward repetition. After reproducing the interesting results discovered by chance on his own body, the child tries sooner or later to conserve also those which he obtains when his action bears on the external environment. It is this very simple transition which determines the appearence of "secondary" reactions; accordingly it may be seen how they are related to the "primary". . . . We can call the circular reactions of the second stage "primary". Their character consist in simple organic movements centered on themselves (with or without intercoordination) and not destined to maintain a result produced in the external environment. So it is that the child grasps for the sake of grasping, sucking, or looking, but not yet in order to swing to and from, to rub or to reproduce sounds . . . On the other hand, in the circular reactions which we shall call "secondary" and which characterize the present stage, the movements are centered on a result produced on the external environment and the sole aim of the action is to maintain this result. (Piaget, 1974, p. 154–157)

This distinction is, as we have seen in our longitudinal studies, very important and it can be extended to schemata in general, independently of whether they are repeated or not, that is, of whether or not they are circular reactions. In this sense we can speak of primary and secondary schemata. Both types are directed to objects but their effects are different. Primary schemata imply a direct assimilation of the object to the body, as in grasping an object or mouthing it. Their knowledge contribution consists in the effect that the object has on the body part, through whatever perceptual channel is involved (kinesthetic, tactile, visual, etc.). This translates into the kinesthetic, tactile, visual image, or combination thereof ("with or without intercoordination," as specified in the quotation above) of the relevant object properties. In secondary schemata the object acted upon produces some "external" result; that is, besides the direct effect on the body, as in primary schemata, it produces an effect on some other "object," *sensu lato*, as when an object is banged onto the ground or it is put in motion. The specific knowledge contribution of secondary schemata consists of this second type of effect. This is perceived through the same channels, which in this case act only as a medium, while in primary schemata they are, so to speak, both the medium and the content: in the first case, they "register" a modification induced on the external world, in the second they "register" a modification induced on themselves. In other words, in primary schemata knowledge is derived from the interaction between object and self, while in secondary schemata from the interaction between object and "object."

Though this is seldom appreciated in conscious adult human cognition, where one's own body tends to be objectively cognized as any other physical object, knowledge constructed from these two sources is quite distinct. Just to give an example, one can take the observations of Kohler on chimpanzees struggling with problems of equilibrium, quoted above in chap. 10. Though we speak in both cases of "balance," "balancing," "equilibrium," etc., the chimpanzee "knows" perfectly well how to balance a pole or a box with respect to his body (holding it in equilibrium, or keeping in equilibrium both himself and the object), but he does not "know" at all how to balance an object relative to another object and fails in the most dramatic way. These examples could be multiplied *ad libitum*. At a more elementary level, the condition that a physical contact must be established between causal agent and target object might not be understood at all, as we have seen in the stick problem, when both the causal agent and the target object are independent objects (the stick and the reward), even if the same subject understands perfectly well the same condition when the causal agent is its own body (its own hand and the stick: no animal fails to grasp the stick in order to act on it).

With respect to the distinction between logical and physical cognition considered in chapter 6, while both these types of knowledge pertain to the physical domain, it is obvious that only the second, the one derived from object-object interactions, leads to a generalized and objective construction of physical relations, and, in fact, it can also "translate" and incorporate the first one, which remains limited to the "special" case of object-self relations.

In the analysis of object manipulations which will follow, we will therefore retain this fundamental distinction between primary and secondary schemata of manipulation, and detail it with further subdivisions.

In this theoretical and knowledge-oriented basis, our account will be also considerably different from previous studies of nonhuman primates, where object manipulation was essentially used as an empirical index of taxonomic distance among species (see, for example, Parker, 1974a, 1974b; Torigoe, 1985).

DATA ANALYSIS

The general transcription was scanned extracting and coding each manipulative event performed on the objects of the experimental set offered. Particular attention was paid to incorporate all the possible consequences that action on an object produced on the surronding environment. Passive dropping (i.e. letting go an object) was not included. While a certain amount of arbitrariness in identifyng, equaling as instances of the same class, and, especially, segmenting manipulative events cannot be avoided, the operational criteria followed were kept strictly constant across subjects and ages. Comparative results should thus be as free of bias as possible.

RESULTS

A total of 40,746 manipulative events were recorded and classified into 36 different action schemata (a. s.). Table 1 reports their distribution among the 5 age/species classes (non-obvious schemata labels are explained in Table 3 below).

As it can be seen, 29 of the 36 a. s. are common to both macaques and cebus, while 6 are found only in cebus (lift to drop, tap mouth on, spring release, hit, line up, stop moving objects) and 1 (scratch) only in macaques. Though the types of a. s. performed are largely common to both species, their relative frequency of occurrence is not. These frequencies covary on the basis of the subjects' species more than on the basis of the subjects' age. Table 2 reports the correlation matrix built on the frequencies of the 29 common a. s.: it is evident how extremely high values of r class together

TABLE 1.
Action Schemata Frequencies Expressed as Percent Values Over the
Total Number of Schemata in Each Subject Class.

	Ceb16	Ceb36	Ceb48	Mac22	Mac34	Tot Ceb	Tot Mac	Total
Mouth	10.89	9.31	10.12	11.08	11.73	10.06	11.33	10.54
Turn ar./up/down	8.74	11.62	8.29	9.45	7.45	9.36	8.69	9.11
Hold in hand	7.47	6.80	7.19	10.44	12.23	7 14	11 12	8.64
Bite	5.62	2.56	3.22	14.07	18.82	3.57	15.88	8.20
Bang	8.81	9.04	15.68	.14	0	12.19	.08	7.64
Away to look	3.31	4.26	4.78	11.50	13.36	4.29	12.21	7.27
Put in contact	5.03	8.57	12.79	2.59	3.25	9.80	2.84	7.19
Slide	11.72	8.65	8.28	1.59	1.27	9.17	1.47	6.28
Hold in mouth	4.35	2.70	2.88	10.49	10.17	3.16	10.37	5.87
Put on ground	5.34	4.51	3.57	2.08	2.11	4.25	2.09	3.44
Touch with hand	1.75	1.31	72	6.08	4.47	1.12	5.47	2.76
Lift	2.81	3.52	3.64	1.90	17	3.42	1.24	2.60
Touch with mouth	5.24	3.16	3.41	.82	.36	3.75	.65	2.58
Press against	5.06	1.60	1.55	1.94	2.43	2.36	2.13	2.28
Slide in mouth	1.20	.34	.33	5.24	4.88	.53	5.10	2.25
Tap hand on	1.02	6.03	1.66	1.06	.03	2.78	.67	1.99
Bring to eyes	.43	2.53	1.77	1.23	74	1.69	1.04	1.45
Rub	1.84	1.92	2.40	.27	.10	2.13	.21	1.41
Break	.02	1.25	1.23	2.14	1.82	.96	2.02	1.36
Put hand in	1.93	1.57	.91	1.56	.53	1.33	1.17	1.27
Lift to drop	2.38	1.78	.77	0	0	1.43	0	.89
Slide across	1.01	.54	.36	1 13	.91	.56	1.05	74
Roll	.47	.83	1.26	.06	.05	.96	.06	.62
Put into (obj.)	.24	1.08	.46	.52	.57	.59	.54	.57
Hit	73	1 15	78	0	0	.88	0	.55
Lick	.45	18	13	.57	1.25	.22	.83	.45
Put mouth in	73	.95	.50	.05	.02	.68	.04	.44
Smell	0	14	.01	71	.99	.04	.82	.33
Tap mouth on	.28	1.37	0	0	0	.46	0	.29
Swing	17	.05	.21	.56	19	16	.42	.26
Throw	.07	15	.62	.03	.03	.36	.03	.24
Spring release	.66	.09	.37	0	0	.36	0	.22
Rub/scratch h on	.05	.07	.02	.55	.09	.04	.37	17
Line up	0	.23	.07	0	0	.10	0	.06
Stop (mov. obj.)	.17	14	.02	0	0	.09	0	.06
Scratch	0	0	0	14	0	0	.08	.03
Total (N)	5766	7378	12293	9469	5840	25437	15309	40746

TABLE 2.
Correlations of Common Schemata Frequencies
Among the 5 Age/Species Classes of Subjects.

	Ceb16	Ceb36	Ceb48	Mac22	Mac34
Ceb16	—	.86	.82	.48	.45
Ceb36	.86	—	.89	.38	.32
Ceb48	.82	.89	—	.29	.28
Mac22	.48	.38	.29	—	.98
Mac34	.45	.32	.28	.98	—

across ages cebus on the one side and macaques on the other side, while all correlation values between different species remain much lower.

Inspection of the last three columns of Table 1 reveals that a few a. s. account for the vast majority of manipulative acts performed. In both species there appears to be a gap between the first six most frequent schemata and the remaining. In macaques, these first 6 (Mouth [an object], Turn [an object] around, up or down, Hold [an object] in hand, Bite [an object], Away [to distance an object] to look [at it], Hold [an object] in mouth), account for 70% of all the events. In cebus, the first 6 most frequent schemata (Mouth, Turn around/up/down, Hold in hand, Bang [an object onto a surface or another object], Put in contact [an object with another object], Slide [an object on a surface or another object]) account for 58% of all the events.

Both the higher number of schemata-types presented (35 vs. 30) and their lower "compactness" (58% vs. 70% of total coverage for the first six schemata) argue for a more discriminative and varied style of manipulation in cebus.

Cebus style of manipulation differs from that of macaques on another dimension too: the frequency of a. s. performed in the same unit of time. These differences are plotted in Fig. 1 in terms of a. s. performed per minute. At all ages, macaques perform less a. s. than cebus in the same amount of time. Furthermore, their rate does not increase with age, while it nearly doubles in cebus, dramatically amplifing this difference. In the same unit of time, cebus have the opportunity of extracting more information from the objects they are interacting with.

Already from this initial generic comparison a sharp difference emerges by simply looking at the nature of the six most frequent a. s. in the two species. Three of them are common to the two species, while three differ already at this high frequency of occurrence. In cebus, these three are Bang, Put in contact and Slide. All three belong to the category of "secondary" schemata, as defined above: they all imply an interaction between the object acted upon and another object. In macaques, on the other hand,

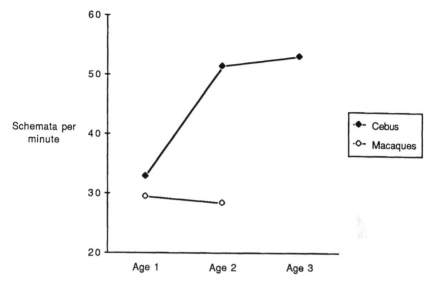

FIG. 1   Mean frequencies of action schemata per minute performed by the two species at each age.

the three non-shared schemata are Bite, Away to look and Hold in mouth: none of them is a secondary schema and two of them (Bite, Hold in mouth) are among the simplest of primary schemata, being performed through the mouth.

In order to further this type of analysis we grouped a. s. in categories, according to the criteria discussed above in the introduction.

Beside the fundamental distinction between primary and secondary schemata, we divided primary schemata between those where the direct assimilation of the object takes place through the mouth (labeled "primary mouth") and those where it takes place through the hand(s) (labeled "primary hand"). Exploration of objects through mouth is more primitive, i.e. capable of less articulation and differentiation, than that through hand. To these three categories we added a fourth one, labeled "visual exploration", comprising those a. s. whose only effect is that of changing the object orientation and/or its perspective view while the subject is looking at it. These schemata cannot be classed, strictly speaking, with either primary or secondary schemata. Their main effect is to allow the comparative inspection of the different parts of an object and to elaborate its internal differentiation. In this respect, they tend to be more informative, so to speak, than directly assimilating primary schemata, but at the same time do not investigate properties of physical interactions as secondary schemata.

Table 3 shows the distribution of all 36 a. s. into the four categories.

TABLE 3.
Groupings of Action Schemata.

Primary mouth	1) Mouth (bring to); 2) bite; 3) lick; 4) smell; 5) hold in mouth; 6) slide in mouth; 7) touch with mouth, 8) put mouth inside; 9) tap with mouth on top.
Primary hand	1) Hold in hand; 2) lift; 3) put on ground[a]; 4) touch with hand; 5) put hand inside; 6) rub hand on top; 7) tap hand on top; 8) scratch.
Visual expl.	1) Bring to eyes; 2) draw away to look at; 3) turn around upside down while looking.
Secondary	1) Press against substrate; 2) slide; 3) roll; 4) release in a springlike fashion[b]; 5) rub against surface; 6) bang; 7) put in contact with another object; 8) slide across[c]; 9) put an object into (other objects); 10) stop (a moving object); 11) hit; 12) swing; 13) throw; 14) lift to drop; 15) break; 16) line up.

[a]: Not including passive letting go or dropping.
[b]: Lifts and releases one end of the object while pressing the other end against ground.
[c]: Slides the object across a hole, the wire-mesh, etc.

Object categorization

First of all let's consider whether and how different objects are categorized in terms of the schemata applied to them. Tables 4 and 5 show, for cebus and macaque respectively, the frequencies and proportions of the four categories of schemata for each of the four different types of objects presented to manipulation. Both species sharply distinguish the four objects from one another, as it can be seen for the widely different distribution of schemata-types characterizing their manipulation (Table 4: chi square = 1039, df = 9, p < .0001; Table 5: chi square = 385, df = 9, p < .0001). Data were collapsed across age classes since there was no significant difference in the distinguishing pattern: objects were manipulated in a closely similar fashion in all age subgroups.

In both species, cups were manipulated more than any other object, and were followed, in order, by sticks, crosses and rings. However, the way in which objects were differentiated greatly differed in the two species. Figures 2 and 3 plot the data of Tables 4 and 5, respectively, as percent deviations from mean values.

In cebus Visual exploration and Secondary schemata show a much wider excursion than Primary schemata, accounting for most of the differences among objects. In Macaques objects are mainly distinguished through the opposing pattern of Primary mouth and Visual exploration schemata.

Objects emerging as maximally different between each other were cups and sticks: cups are marked by the higher frequencies of Visual exploration schemata in both species. On the other hand cups show very low frequencies of Secondary schemata in cebus and of Primary mouth schemata in ma-

Table 4.
Cebus' Frequencies and Proportions (in parentheses) of
Schemata Categories in Manipulating Different Objects.

	Cups	Rings	Crosses	Sticks	Total
Primary mouth	1801	974	1127	1285	5187
	(22.8)	(19.3)	(21.5)	(21.6)	(21.5)
Primary hand	1716	922	1037	1238	4913
	(21.7)	(18.2)	(19.7)	(20.8)	(20.3)
Visual explor.	1803	858	624	361	3646
	(22.8)	(17.0)	(11.9)	(6.1)	(15.1)
Secondary	2580	2303	2465	3059	10407
	(32.7)	(45.5)	(46.9)	(51.5)	(43.1)
TOTAL	7900	5057	5253	5943	24153
	(100)	(100)	(100)	(100)	(100)

Table 5.
Macaques' Frequencies and Proportions (in parentheses) of
Schemata Categories in Manipulating Different Objects.

	Cups	Rings	Crosses	Sticks	Total
Primary mouth	2019	551	1465	2120	6155
	(37.8)	(41.1)	(49.2)	(51.2)	(44.6)
Primary hand	1272	353	662	811	3108
	(23.8)	(26.3)	(22.2)	(19.6)	(22.5)
Visual explor	1580	283	486	728	3077
	(29.6)	(21 1)	(16.3)	(17.6)	(22.3)
Secondary	476	145	367	483	1471
	(8.9)	(10.8)	(12.3)	(11.7)	(10.6)
TOTAL	5347	1342	2980	4142	13811
	(100)	(100)	(100)	(100)	(100)

caques. Sticks, on the contrary, show the highest frequencies of Secondary
schemata in cebus and of Primary mouth schemata in macaques, while in
both species they are marked by an extremely low level of Visual Exploration schemata.

Differences between species

Differences between cebus and macaques in categorizing objects through
manipulation, directly reflect their different patterns of dealing with the
objects they act upon.

Table 6 summarizes the frequencies and proportions of occurrence of

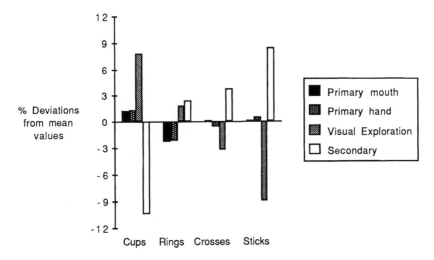

FIG. 2.   Distribution of action schemata categories for each object type in cebus. Values reported are percent deviations from mean values.

the four schemata categories in the five age/species groups. As it can be seen, cebus production of secondary schemata is on the average 4 times higher than macaques, while the most elementary primary schemata performed through the mouth are half or less than half those of macaques.

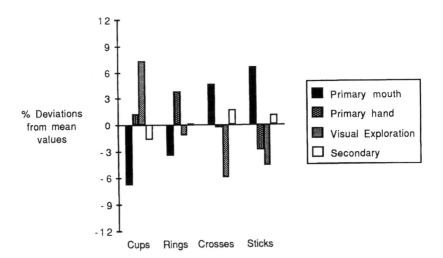

FIG. 3.   Distribution of action schemata categories for each object type in macaques. Values reported are percent deviations from mean values.

Table 6.
Absolute Frequencies and Percent Values (in parentheses) of Each
Schemata Category in the 5 Species/Age Classes.

	Primary Mouth	Primary Hand	Visual Explorat.	Secondary	Total
Cebus 16	1658	1175	720	2213	5766
	(28.7)	(20.4)	(12.5)	(38.4)	(100)
Cebus 36	1527	1758	1358	2735	7378
	(20.7)	(23.8)	(18.4)	(37.1)	(100)
Cebus 48	2531	2177	1824	5761	12293
	(20.6)	(17.7)	(14.8)	(46.9)	(100)
Macaques 22	4074	2255	2100	1040	9469
	(43.0)	(23.8)	(22.2)	(11.0)	(100)
Macaques 34	2813	1146	1258	623	5840
	(48.2)	(19.6)	(21.5)	(10.7)	(100)

This difference was already evident in the comparison between the total values of Tables 4 and 5. The data of Table 6, however, add an interesting developmental dimension: this difference tends to increase with age. Fig. 4 plots the proportions of primary mouth and secondary schemata in the two species at each age.

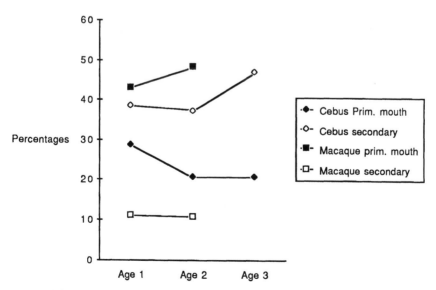

FIG. 4. Percent values of primary-mouth and secondary schemata performed by the two species at each ages.

Primary mouth schemata increase in macaque and decrease in cebus, while secondary schemata show the opposite trend. The two species tend to reinforce, with age, their preferred pattern of dealing with objects.

Given the crucial role of secondary schemata in constructing the physical domain of knowledge, let's analyze in more details their specific contribution in the two species. A subset of these schemata have the property of making more salient the fundamental relationship of cause and effect. This happens because they share three crucial features facilitating the observation and experimentation of the causal action:

1) Their effect on substrates is clearly perceptible;
2) Their effect can vary along a continuous dimension and systematically covaries with the action: for example, the noise produced by a single act of banging is directly proportional to the strength of banging;
3) They result in a modification of the state of the object acted upon: for example, an object can pass from a static to a dynamic state or viceversa by making it roll or by stopping it while moving.

We shall call these schemata "strictly causal secondary schemata", and include in this category the a. s. of stopping (mov. obj.), springlike releasing, throwing, lifting to drop, hitting, rolling, rubbing and banging objects.

In the youngest cebus these schemata are already 18 times as frequent as in the youngest macaques (872 occurrences vs. 48), and such a wide difference becomes larger with increasing age. They increase regularly in cebus from Age 1 to Age 2 (1,114 occurrences), to Age 3 (2,693), while they virtually disappear in Macaques, passing from 48 occurrences at Age 1 to 11 at Age 2. This trend is clearly evident, whether considering the absolute frequencies of strictly causal secondary schemata (Fig. 5) or their proportion over all secondary schemata (Fig. 6).

Thus, not only are secondary schemata more widely employed by cebus than by macaque, but this difference is also much larger in the case of those secondary schemata that offer more opportunities to construct the properties of physical causality.

As a further check on this important difference that directly affects the physical domain of cognition, a further analysis of causal interactions was conducted. This time we looked at the effects action might produce on the interactions between two independent objects. Strictly speaking, this analysis concerns compositions (according to the criteria that will be specified in chapters 13 and 14) rather than manipulations, since in these cases (at least) two objects are combined together. Nevertheless, in view of the relevance of these compositions to the topic of physical cognition, it will be presented here.

Every time at least two objects are actively put in contact with each other, it is possible to distinguish between two basic kinds of relations:

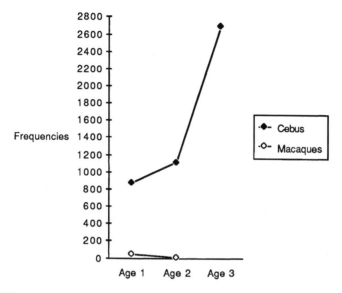

FIG. 5. Absolute frequencies of causal secondary schemata performed by the two species at each age.

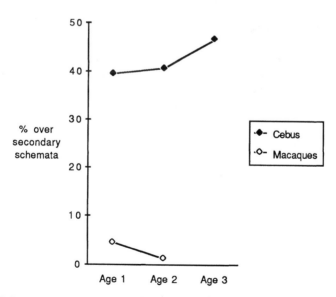

FIG. 6. Proportion of causal secondary schemata over total number of secondary schemata performed by the two species at each age.

Table 7.
Absolute Frequencies of Causal and Spatial
Compositions in the 5 Species/Age Classes.

	Causal Compositions	Spatial Compositions
Cebus 16	156	159
Cebus 36	255	405
Cebus 48	802	734
Macaque 22	11	272
Macaque 34	0	338

1) When object A produces a perceivable effect on object B (noise, displacement, tilting, breaking, etc.), **causal** relations between objects are prevalent. The resulting compositions will be labelled "causal compositions".

2) When object A is placed on the top of, or on the side of, or into object B, **spatial** relations between objects are prevalent. The resulting compositions will be labelled "spatial compositions".

In case (1) object A is used as a physical intermediary to act on object B, and it induces a change of state of B (e.g.: the silent object becomes noisy, the stabile becomes mobile, the flat becomes tilted, the whole becomes partitioned, etc.); In case (2), instead, no such change is produced: the state of object B is not transformed by, and is not dependent on the kind of action performed on object A.

Table 7 reports absolute frequencies of both types of compositions in the five species/age classes. While the number of spatial compositions produced by the two species is roughly comparable, that of causal compositions shows an incredible difference. Only 11 such compositions were produced by macaques at 22 months and none at all at 34 months. The plots of Fig. 7 (7a: Causal compositions, 7b: Spatial compositions) evidence the developmental split between the two species.

Cebus experiment with both kinds of constructions and regularly increase their numbers with age. Macaques do so with only spatial constructions, and, in fact, by age 34 months do not actively provoke any causal interaction between independent objects.

DISCUSSION

The general thrust of these results appears not only to support but also to provide further evidence in favor of the existence of those cognitive developmental divergences among primate species that emerged in the studies reported in the preceding chapters. Structure of logical cognition, at least

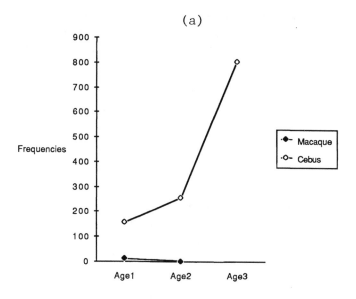

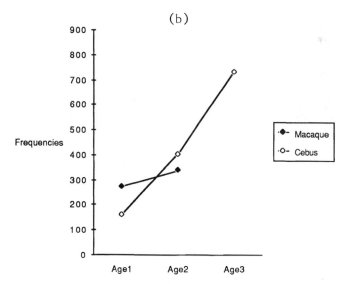

FIG. 7   Frequencies of causal compositions (7a) and spatial compositions (7b) constructed by the two species at each age.

at the elementary level shown by classification through action schemata, is equally developed in the two species of nonhuman primates and is well established in both of them already at the first ages tested. All of our subjects, even at their youngest age, are well beyond the phase in which all discrete objects are fused in a semantically indistinct whole, where they indiscriminately share properties, such as being "mouthable", "touchable", "holdable", "turnable" and so on. Such a phase is typical of human infants at 6 months of age (Langer, 1980). To quote Langer (p. 33): "Subjects' mappings are the overriding determiners of their transactions at any particular time; objects are not". On the contrary, the different types of manipulations that our subjects performed on the different kinds of objects offered appear to be dependent on the internal features of the object acted upon. This is why, for example, both cebus and macaques show their highest frequencies of Visual Exploration schemata with cups. Among the objects presented, cups are characterized by the highest degree of internal differentiation (they have different top and bottom, an inside and an outside, etc.), and their appearance changes greatly when looked at from different perspectives. As a consequence, part-whole relations are far more investigated in cups (for example by continuously rotating them while visually monitoring the prospective transformations they undergo) than other objects.

Sticks, from this point of view, maximally differ from cups. Since they are minimally internally differentiated, only minor differences result by changing the position or perspective of the object. As a consequence, the pattern of manipulation elicited by sticks appears to be opposite to that elicited by cups in both species. Sticks have the lowest frequencies of Visual Exploration schemata, and are instead explored through other modalities (they are predominantly put in relation with other objects or surfaces by cebus and simply mouthed by macaques).

Wide differences characterize, instead, the manipulations of objects of the two species in so far as they pertain to the construction of physical cognition, and more specifically, of causal and dynamic properties of objects and object interactions. From richness in the variety of secondary schemata, to the extensiveness of their application, to the specific probing of a variety of causal effects, to the investigation of cause and effect relations in object-object interactions, macaques consistently function at a level which is far below that of cebus. The more elaborate the physical structure considered (such as the regulation of causal interactions between two independent objects), the wider the difference between the two species, with macaques lagging more and more behind. Furthermore, contrary to cebus, macaques show progress with age in none of these measures: in fact, their trend is either stability or regression.

On the other hand, cebus experimenting with object-object causal relations is akin to that shown by children in similar manipulative situations.

In Langer's (1980) analysis of children compositions, causal compositions account for 35% of composition at 8 months and 47% at 10 months.

These results match perfectly those of the longitudinal studies and of the stick problem. The physical domain of cognition is not being constructed by macaques, not even at the older ages considered in this experiment, while no comparable lack emerges in the logical domain, or certainly not to the same degree. Cebus, on the other hand, show rich and widespread secondary manipulations, thus confirming the initial trend seen toward the end of the longitudinal studies. Though much later in development, in both absolute and comparative terms, they appear to be able to construct at least the fundamental notions of physical cognition.

13 Classification

Giovanna Spinozzi and Francesco Natale
Istituto di Psicologia, C.N.R.
Rome, Italy

INTRODUCTION

Though the study of classification has been one of the cornerstone of cognitive and developmental psychology, it has been traditionally directed to children not younger than 2, i.e. to children already possessing speech and a well developed representational capacity. In fact, some traditional views (most notably, Vygotsky's, 1962) considered the semantic organization of language as more or less essential to the development of classification. But even opposers of such views, as Piaget, did not address (at least, in a systematic research effort) the question of the pre-representational origin and development of classification.

Since the pioneering studies of Ricciuti (1965) and Nelson (1973), a number of investigators have succeeded in pushing back the age at which children's classification abilities could be tested (Starkey, 1981; Sinclair et al., 1982; Forman, 1982; Sugarman, 1983), by analyzing the manipulations of objects spontaneously produced by infants. The broad-scale developmental picture emerging from such studies can be summarized as follows. Evidence for the beginning of classifying activity in the first year of life comes from the differential assimilation of objects to different schemata, i.e., from the two processes of "generalizing assimilation" and "recognitory assimilation". This classification activity, operated only in action, is of the same type as the one we have just considered in the preceding chapter in analyzing the differentiation of object types in terms of their differential manipulatory patterns.

By 6–9 months of age, children begin to form groups of objects, whereby classification can be studied in both their sequential sorting activity and the products thus generated. By the first half of the second year, both these spontaneous responses become organized according to the class membership of the objects, with serial manipulation of identical objects usually preceding spatial grouping of the same objects.

In order to carry out a comparison of the classificatory activity of human and nonhuman primates, we can therefore examine the free manipulations of objects performed by our subjects, extract all the cases in which they spontaneously select and group objects and determine the class properties of both their sequential choices and the resulting compositions, in relation to the class membership of the objects that were offered for manipulation.

The interpretation of the results obtained in this type of research has, however, been subjected to question.

> This evidence for an awareness that "there are two (several, etc.) things that are the same" is not as compelling as it might seem. As already suggested, children who sequentially contact (and even group) same-class objects might not be comparing those objects at all (a comparison of objects is implied by the notion that "there are two things that are the same"). They simply could be picking up what stands out, and those things that stand out happen to belong to one class. . . . [in fact]. . . . children between 1 and 1 1/2 are more likely to select objects from only one class than from more than one in any given episode of (temporal or spatial) class grouping, despite the availability of two or more classes. . . . children could be selecting same-class objects on the basis of salience and not on the basis of interobject comparison. . . . The case is more compelling by 24 months, when the frequency of groupings of two or more classes increases . . . A child who sets aside two class groups (e.g., a group of dolls and a group of blocks, rather than only dolls) could have both classes in mind simultaneously. (Sugarman, 1983, pp.8–9)

Even grouping objects into two (or more) consistent classes would, however, not provide stringent evidence that the subject is not operating on the basis of object saliency:

> Young children could in principle generate groupings of several classes by repeatedly looking for an object like the one they previously placed into the array, that is, by considering only one kind of thing at a time. . . . a child might find a square, look for something like it, find another one, look for something like that, etc. . . . at some later point the child might encounter a circle, look for something like it, find another circle, etc. This strategy would have a high chance of success *if the objects could only be divided in one way, as in the studies we have reviewed.* (Sugarman, 1983, p.10, italics added)

This is, in fact, the very way in which Inhelder and Piaget (1964, chap.1) interpreted the first successive class groupings ("figural collections") they found in their subjects. Ideally, one would like to see the *simultaneous* construction of two separate and coherent classes of objects, something, however, that children begin to produce only around 2 years of age.

One of the ways to make a stronger case for actual classification, even if only one class of objects is constructed, would be to avoid using in the test only the condition described in the italicized portion of the quote above, that is to present only two mutually exclusive classes of test objects. Class membership of objects could be rendered on the one side more ambiguous to decide upon and on the other side impossible to perform on strictly identical objects.

Our design, following Langer (1980), did exactly that. Beside presenting two mutually exclusive classes of objects, which, as seen in chap. 11, corresponds to our Disjoint class condition, we also presented two conditions in which class membership is ambiguous. In the Additive class condition, the two classes of objects presented are not mutually exclusive but share one of the two properties defining them. Thus, while two objects belonging to the same class are still identical, two objects belonging to two different classes are half similar, and, furthermore, this similarity can be either in form or in material (the set of 6 objects is, for example, composed of 3 wood cups and 3 plastic cups). Thus, in order to classify, the subject should resist putting together object that are, for example, only of the same shape, and must pay attention to both attributes defining the two classes. In the Multiplicative class condition, the strategy of putting together successive identical objects cannot work at all, since in this condition no two objects are identical. In order to classify, the subject must put together objects that share one property (either form or material) but are different in the other property. Furthermore, the same type of physical combination of objects, which represents class consistency in the Multiplicative condition, represents class inconsistency in the Additive condition.

In this connection, it should be remembered that there are two systematic behaviors to take into account in order to evaluate classification, when using the technique of spontaneous sorting and grouping. If objects are selected and put together by taking into account their class membership, this can give rise, in principle, to two complementary strategies: either searching for maximally similar objects or searching for maximally different objects, in each condition. Both these strategies are systematic from the point of view of classification, because they both imply taking into account the classes into which the presented objects can be divided, and are opposed to selecting and putting together objects disregarding their class membership, where no systematicity in the selections or groups can be found.

Appropriate tests should thus evaluate both these possibilities. Former studies (with the exception of Langer, 1980) have instead failed to take into account the possibility of selecting and grouping objects in a systematic class-inconsistent way. This possibility becomes all the more important to consider when the test embodies ambiguous class conditions, as in our Additive and Multiplicative conditions. In these conditions, in fact, as we just saw, the very same selecting and grouping regularities might result in class consistency in one condition and class inconsistency in the other.

In conclusion, it seems to us that the design of our test situations coupled with appropriate analysis of the results obtained, should make possible to gather enough converging evidence to determine whether and to what extent real classificatory abilities are present in our subjects, even when the construction activity is confined to one class at a time, which is by far the predominant pattern of our subjects' composing (see the next chapter).

In the Discussion section we will compare, in some detail, the results of this study with those obtained by Langer (1980, 1986) on human infants from 6 to 24 months, which, as noted in the introduction, are the most homogeneous in terms of data collection methodology and data analysis.

DATA ANALYSIS

Since the purpose of this analysis was to determine to what extent the spontaneous groupings of objects produced by the subjects' manipulatory activity embedded properties relevant to classification, transcriptions were scanned to extract all cases in which two or more objects had been put together and to code the identity of the objects put together as well as the order in which they were manipulated.

Two criteria had to be conjointly satisfied in order to consider a group of objects as a composition: (1) the objects had to be either in contact or in close spatial proximity to each other, where "close spatial proximity" means within a maximum distance of 10 cm from each other; (2) at least one of the objects had to have been actively manipulated to produce the composition.

A composition was demarcated from another occurring in temporal succession only if at least one of the following two criteria was satisfied: (1) the spatial position of (at least) one object of the composition was modified in such a way as to change its relations of adjacency with (at least one of) the other objects; (2) one (or more) object(s) was (were) added to the composition.

All compositions identified according to these criteria were subjected to two types of analysis. *Object Composing*, in which the class properties of each resulting composition were considered; *Object Selecting*, in which the class properties of the succession of objects manipulated to form each composition (or "order" of manipulation, for brevity) were considered. In *Object Composing* a composition was scored as "class consistent", or "unmixed", if all the objects making up the composition belonged to only one of the classes presented in each condition. If instead the objects belonged to different classes the composition was scored as "class inconsistent", or "mixed".

First of all, it should be noticed that this distinction can be applied to compositions of two or three objects only. In fact, since in each condition presented there is a maximum of three objects belonging to one and the same class, it is obvious that compositions of more than three objects must necessarily be "mixed". Therefore, in this analysis only compositions of two and three objects were taken into account.

Second, and most importantly, class-consistent and class-inconsistent compositions correspond to different products in the different class conditions. In the Additive and Disjoint conditions class consistency means composing identical objects, i.e., of *both* the same form *and* the same material, but class inconsistency in the Disjoint condition means composing (at least two) objects completely different, i.e. in both form and material, while in the Additive condition it means composing (at least two) objects partly similar, i.e. of either the same form or the same material. On the other hand, class consistency in the Multiplicative condition means composig objects only similar, i.e., of *either* the same form *or* the same material, since in this condition there are no identical objects, while class inconsistency means composing (at least two) objects completely different, i.e. in both form and material. The following summarizes which types of object groupings (where "different" means two objects that are completely different, i.e. do not share any of the two properties) correspond to consistency and inconsistency in the different class conditions.

	Class Consistency	**Class Inconsistency**
Additive	Identical	Similar
Disjoint	Identical	Different
Multiplicative	Similar	Different

Furthermore, in the Multiplicative condition, and only in it, composing in a class consistent way offers in every presentation two possibilities: the subject can indifferently choose, as dimension of similarity, the property embodied in 3 classes of 2 objects each or that embodied in 2 classes of 3 objects each. For example, if there are a plastic and a wood ring, a plastic and a wood cross, and a plastic and a wood stick, the subject can choose form as a criterion of similarity, in which case it can form one, two or three classes (rings, crosses or sticks) of two objects each, or it can choose material, in which case it can form one or two classes of up to three objects each (plastic objects or wood objects). This option offers the possibility of judging rigidity vs. flexibility in class grouping of the same physical objects.

In Object Selecting, since class properties are judged over the active choice of objects effected by the subject in constructing compositions, only those compositions where at least two objects were actively manipulated can be considered. On the other hand, since there are at most three objects belonging to the same class in each condition, only compositions constructed with at most three successive object choices can be considered in this analysis.

Criteria defining class-consistent vs. class-inconsistent selections in the different class conditions are the same as in Object Composing.

Notice, however, that compositions taken into account for analysing Object Selecting can be completely different from those analysed in Object Composing. For example, all two-object compositions where only one object has been manipulated (by putting it in contact with or close to another object which has not been touched) will be scored for Object Composing but not for Object Selecting. If two objects are successively manipulated and combined with a third, untouched object, two compositions will be scored for Object Composing (a two- and a three-object composition), but only a two-object selection will be scored for Object Selecting. Furthermore, if in this last case, the two objects selected belong both to one of the two classes but the untouched object belongs to the other class, then the two compositions will be both scored as "mixed" for Object Composing, while the two-object selection will be scored as "unmixed". For example, if a plastic ring is put next to a wood cross, a "mixed" composition of two objects will be scored and nothing will be scored for Object Selecting because only one object has been manipulated; if a second plastic ring is then put next to the former two objects, then another mixed composition, a three-object one, will be scored, but this time also an "unmixed" two-object selection will be scored, because the subject has manipulated two identical objects in forming a three-object composition. Finally, Object Selecting might regard also four-, five- and six-object resulting compositions, which are not taken into account in Object Composing, when these are formed by adding only two or three objects to an already existing group of objects.

In short, the two analysis are largely independent and might reflect different types of classificatory behaviors, one centered on the product of the construction and one on the process of construction.

TABLE 1.
Number and Proportion of Two, Three and Four-or-More-Object
Compositions by 22 and 34 Month-Old Macaques.

	Set Size						Total
	Two		Three		Four-or-More		
	N	%	*N*	%	*N*	%	*N*
22 months	369	85.4	43	9.9	20	4.6	432
34 months	297	77.9	54	14.2	30	7.9	381

RESULTS

Macaques Object Composing

Set size

Table 1 reports the number and proportion of compositions generated by Macaques at 22 and 34 months according to their size.

Most compositions are of the minimal possible size, 2 objects. Even at the most advanced age, 34 months, over three quarters of all compositions produced comprise two objects only.

There seems to be, however, a coherent developmental trend. Compositions of the minimal size decrease with increasing age, while those with three, four or more objects increase. Total number of compositions produced does not, however, increase with age: on the contrary, it shows, if anything, a slight tendency to decrease.

Class properties of compositions

Table 2 reports the number of class-consistent ("unmixed") vs. class-inconsistent ("mixed") compositions produced in each class condition, and separately for two-object and three-object compositions, at 22 months. Table 3 reports the same data for the 34-month age.

In order to determine whether compositions generated in each class condition are systematically class consistent, or systematically class inconsistent (or, obviously, neither, i.e., objects are put together randomly with respect to their class membership), statistical comparison of each pair of unmixed vs. mixed total was effected by means of two-tailed Binomial tests. In the Additive and Disjoint conditions the expected probabilities of composing two objects in a class-consistent vs. class-inconsistent way are .4 vs. .6. In the Multiplicative condition these values are reversed: .4 for class-inconsistent and .6 for class-consistent two-object compositions. In three-object compositions, the expected probabilities of class consistency

TABLE 2.
Number of Unmixed and Mixed Two and Three-Object
Compositions by 22 Month-Old Macaques.

	Set Size				
	Two Objects		Three Objects		
	Unmixed	Mixed	Unmixed	Mixed	Total
Additive	[a]58	47	0	6	111
Disjoint	39	[b]108	1	23	171
Multiplicative	[a]83	34	3	10	130

[a]p < .01
[b]p < .001

TABLE 3.
Number of Unmixed and Mixed Two and Three-Object
Compositions by 34 Month-Old Macaques.

	Set Size				
	Two Objects		Three Objects		
	Unmixed	Mixed	Unmixed	Mixed	Total
Additive	[b]75	37	0	[a]43	165
Disjoint	51	55	0	6	112
Multiplicative	[b]64	15	2	3	84

[a]p < .01
[b]p < .001

and class inconsistency are, respectively, .1 and .9 in all three class conditions.

Results of these tests are indicated on the tables: pairs marked by asterisks indicate significant deviations from random distribution in the direction of the type (unmixed or mixed) where the asterisk appears; pairs unmarked indicate no significant deviation from random distribution; pairs where the total number of compositions was 4 or less were not subjected to test and are marked by a cross.

At 22 months two objects are composed in systematic way with respect to their class properties in every condition, but in different ways. In the Additive and Multiplicative condition two-object compositions are systematically class consistent, i.e. the subjects put together objects that are identical (Additive) or similar (Multiplicative) to each other, while in the Disjoint condition they are systematically class inconsistent, i.e., the subjects put together objects that maximally differ from each other.

On the other hand, at the same age, three-object compositions are random with respect to their class properties in all conditions. At 34 months,

TABLE 4.
Number and Proportion of Compositions According to the
Number of Objects Manipulated by 22 and 34 Month-Old
Macaques to Produce Them.

Age	Number of Objects Manipulated				
	1	2	3	4+	Total
22 months					
Number of compositions	113	306	9	4	432
% Compositions	26.2	70.8	2.1	0.9	100
34 months					
Number of compositions	51	292	36	2	381
% Compositions	13.4	76.6	9.4	0.6	100

two-object compositions continue to be systematically class consistent in the Additive and Multiplicative condition, in fact even more strongly so than at 22 months. In the Disjoint condition, instead, where they were systematically class inconsistent, they become random. Three-object compositions continue to be random except that in the Additive condition. Here objects are systematically grouped by difference.

In the Multiplicative condition, where in classifying by similarity a choice could be made between the two properties of form and material, macaques systematically choose material as the common property defining class membership at both ages: at 22 months, of the 83 "unmixed" composition of the Multiplicative condition, 66 were by material and only 17 by form (the difference is significant: $p < .001$); at 34 months, the corresponding figures are 59 by material and 5 by form ($p < .001$).

Macaques Object Selecting

Number of objects selected

Table 4 reports the number and proportion of compositions generated divided according to the number of objects manipulated to make them. At both ages, most compositions, about three quarters of their total, are generated by actively manipulating two objects. On the other hand, the proportion of constructions produced by maneuvering only one object strongly declines and is cut in half in passing from 22 to 34 months. At the same time there is a strong increase in the number of compositions produced through the manipulation of three objects, which are almost nonexistent at 22 months. Their total number stays, however, quite low even at 34 months (36 compositions). Manipulation of four or more objects in

TABLE 5.

Number of Unmixed and Mixed Two and Three-Object Selections by 22 Month-Old Macaques to Produce Compositions.

	Number of Objects Selected				
	Two Objects		Three Objects		
	Unmixed	Mixed	Unmixed	Mixed	Total
Additive	[b]49	34	0	2[c]	85
Disjoint	28	[b]96	1	2[c]	127
Multiplicative	[a]74	25	0	4[c]	103

[a]p < .01
[b]p < .001
[c]Not tested

TABLE 6.

Number of Unmixed and Mixed Two and Three-Object Selections by 34 Month-Old Macaques to Produce Compositions.

	Number of Objects Selected				
	Two Objects		Three Objects		
	Unmixed	Mixed	Unmixed	Mixed	Total
Additive	[b]80	43	5	29	157
Disjoint	[a]52	45	0	1[c]	98
Multiplicative	[b]57	15	0	1[c]	73

[a]p < .01
[b]p < .001
[c]Not tested

constructing compositions is instead practically absent always. As in object composing, the picture is that of a slight quantitative development on a stable background of minimal manipulative activity.

Class properties of Object Selecting

Table 5 reports the number of class-consistent ("unmixed") vs. class-inconsistent ("mixed") successive object selections in generating compositions for each class condition, separately for selections of two and three objects, at 22 months. Table 6 reports the same data for the 34-month age. As in Object Composing, in order to determine whether the successive choices of objects to generate a composition were systematically class consistent, systematically class inconsistent or random with respect to the class membership of the objects selected, statistical comparison of each pair of unmixed vs. mixed total was effected by means of two-tailed Binomial tests. Probabilities of randomly selecting two and three objects in a class-consistent vs. class-inconsistent way in the different class conditions are as in two and three object composing, respectively.

Results at 22 months are in remarkable agreement with those of Object Composing. The selection of two objects is always systematic with respect to the class properties of the objects in all conditions. It is, furthermore, as in two-object composing, systematically class consistent in the Additive and Multiplicative condition and systematically class inconsistent in the Disjoint condition. On the other hand, selections of three objects were so low in number, as noted above, that they were not subject to test.

At 34 months, class-consistent selection in the Additive and Multiplicative conditions is confirmed, as it was in Object Composing. On the other hand, selection becomes systematically class consistent also in the Disjoint condition, while Object Composing became random in this condition. As noted above, at 34 months selections of three objects remained low in number, except that in the Additive condition. In this condition, selection was random, while three object compositions at 34 months were systematically class inconsistent.

Choice of form vs. material as the property defining class membership in the Multiplicative condition matches exactly that of Object Composing: classification is systematically by similarity in material and not in form at both ages. At 22 months, of 74 "unmixed" two-object selections in the Multiplicative condition, 59 were by similarity in material and 15 by similarity in form (p < .001). At 34 months, 54 were by similarity in material and only 3 by similarity in form (p < .001).

Cebus Object Composing

Set size

Table 7 reports number and proportion of compositions produced by cebus at each age according to their size. As it can be seen, Object Composing undergoes a clear developmental progression: the total number of compositions generated increases sharply, nearly doubling at each successive age, and, at the same time, the number of objects composed into each set also increases. Though minimal, two-object compositions represent the majority of compositions at all ages, their number undergoes a regular decrease, which corresponds to a regular increase of the number of three and larger compositions. In fact, at 48 months, three-object compositions are nearly a third of all compositions produced. Larger compositions, instead, though growing, stay relatively low in number and proportion.

TABLE 7.
Number and Proportion of Two, Three and Four-or-More-Object
Compositions by 16, 36 and 48 Month-Old Cebus.

| Age | Set Size | | | | | | Total |
| | Two | | Three | | Four-or-More | | |
	N	%	N	%	N	%	N
16 months	418	81.2	62	12.0	35	6.8	515
36 months	680	73.0	175	18.8	77	8.2	932
48 months	1350	65.8	542	26.4	160	7.8	2052

TABLE 8.
Number of Unmixed and Mixed Two and Three-Object
Compositions by 16 Month-Old Cebus.

| Conditions | Set Size | | | | |
| | Two Objects | | Three objects | | |
	Unmixed	Mixed	Unmixed	Mixed	Total
Additive	40	69	0	21	130
Disjoint	57	83	0	28	168
Multiplicative	ᵃ124	45	0	13	175

ᵃp < .001

Class properties of Composition

Tables 8, 9, and 10 report all occurrences of class-consistent ("unmixed") and class-inconsistent ("mixed") two- and three-object compositions, divided by class condition, generated at each of the ages considered. The same statistical comparison as in macaques' Object Composing (two-tailed Binomial tests of each pair of unmixed and mixed total) was effected.

Results show that at 16 months cebus composing does not take into account class properties of the objects composed: composing is random with respect to the class membership of the objects in all conditions and types of composition except one. In the Multiplicative condition, two objects are combined in a systematic class-consistent way.

At 36 months, all two-object compositions are instead systematic with respect to the class membership of the objects composed. Objects are, however, composed in class-consistent way in the Additive condition and in a class-inconsistent way in the Disjoint and Multiplicative conditions. Compositions of three objects, instead, remain random in all class conditions. At 48 months, the picture becomes highly coherent. With one

TABLE 9.
Number of Unmixed and Mixed Two and Three-Object
Compositions by 36 Month-Old Cebus

	Set Size				
	Two Objects		Three Objects		
Conditions	Unmixed	Mixed	Unmixed	Mixed	Total
Additive	ª116	114	11	57	298
Disjoint	73	ª174	10	54	311
Multiplicative	101	ª102	7	36	246

ªp < .001

TABLE 10.
Number of Unmixed and Mixed Two and Three-Object
Compositions by 48 Month-Old Cebus

	Set Size				
	Two Objects		Three Objects		
Conditions	Unmixed	Mixed	Unmixed	Mixed	Total
Additive	ª234	150	ª67	75	526
Disjoint	ª332	129	ª86	121	668
Multiplicative	ª372	133	26	177	708

ªp < .001

exception only, all compositions in all conditions are systematically class consistent. Only three-object compositions in the Multiplicative condition remain random.

Choice between form and material as criterion of class membership in the Multiplicative condition is not systematic as in macaques. At 16 months, there is practically an equal number of unmixed compositions with similarity in form and similarity in material: 64 by form and 60 by material. At 48 months, similarity by form is instead predominant: 254 unmixed compositions are similar by form and 118 by material (p < .001)

Cebus Object Selecting

Number of objects selected

Table 11 reports number and proportion of compositions generated at each age divided according to the number of objects manipulated to make them. At 16 months about one half of the composition are generated by manipulating one object only, while the other half is generated by manipulating two objects. These figures change only slightly at 36 months: there is a

TABLE 11.
Number and Proportion of Compositions According to the
Number of Objects Manipulated by 16, 36 and 48 Month-Old
Cebus to Produce Them.

	Number of Objects Manipulated				
	1	2	3	4+	Total
16 months					
Number of compositions	270	236	8	1	515
% Compositions	52.4	45.8	1.6	0.2	100
36 months					
Number of compositions	406	490	30	6	932
% Compositions	43.6	52.6	3.2	0.6	100
48 months					
Number of compositions	839	1105	184	5	2133
% Compositions	39.3	51.8	8.6	0.3	100

decrease in manipulations of one object only and a parallel increase in manipulations of two objects. Manipulations of three objects begin to appear at this age, but their number is very low. At 48 months two-object selections continue to be about one half of all selections. On the other hand, one-object selection decrease further and three-object selection become well established. Four- or-more object selections are instead practically absent at all ages. There seems, therefore, to be a moderate but consistent development in the number of objects manipulated to produce compositions.

Class properties of Object Selecting

Tables 12, 13 and 14 report, respectively for each age point, the number of class-consistent ("unmixed") vs. class-inconsistent ("mixed") object selections in generating compositions for each class condition and for each number of manipulations. Statistical comparison of each pair was made in the same way as in macaques' Object Selecting. Both at 16 and at 36 months all selections where the analysis was possible (at least 5 compositions generated) were random with respect to the class membership of the objects selected. At 48 months this picture changes sharply and becomes almost identical to that of Object Composing at the same age. In all conditions and types, selections become coherently class consistent. Thus, at this age systematic classification by similarity marks all cebus composing activity.

Choice between form and material as criterion for class membership of objects in the Multiplicative condition is again not rigid. In two-object selecting, there are 219 selections by identity in form and 98 by identity

TABLE 12.
Number of Unmixed and Mixed Two and Three-Object
Selections by 16 Month-Old Cebus to Produce Compositions.

| | Number of Objects Selected | | | | |
| | 2 | | 3 | | |
	Unmixed	Mixed	Unmixed	Mixed	Total
Additive	20	37	0	1[a]	58
Disjoint	39	55	0	5	99
Multiplicative	59	26	0	2[a]	87

[a]Not tested

TABLE 13.
Number of Unmixed and Mixed Two and Three-Object
Selections by 36 Month-Old Cebus to Produce Compositions

| | Number of Objects Selected | | | | |
| | 2 | | 3 | | |
	Unmixed	Mixed	Unmixed	Mixed	Total
Additive	72	101	0	14	187
Disjoint	57	119	0	11	187
Multiplicative	78	63	1	4	146

in material, the difference being significant ($p < .001$), but in three-object selecting no such preference is found: 7 selections are by identity in form and 8 by identity in material.

DISCUSSION

Fig. 1 summarizes macaques' overall extent of classificatory behavior, giving it a graphic representation. At each of the two ages there is a total of 6 possibilities of manifesting systematic classification in Object Composing (COMP) and 6 in Object Selecting (SELC): 2 possibilities for each of the three class conditions, one in compositions and selections with 2 objects and one with 3 objects.

As it can be seen, at both ages and in both composing and selecting, macaques classify in half of these possibilities (3 out of 6). Fig. 1 shows also the distribution of classification in the two-object and three-object cases. With one exception only (in Object Composing at 34 months), classification is performed only in the two-object cases.

TABLE 14.
Number of Unmixed and Mixed Two and Three-Object
Selections by 48 Month-Old Cebus to Produce Compositions

	Number of Objects Selected				
	2		3		
	Unmixed	Mixed	Unmixed	Mixed	Total
Additive	[a]178	105	[a]50	26	363
Disjoint	[a]301	100	[a]37	20	458
Multiplicative	[a]317	104	[a]15	32	468

[a]$p < .001$

Thus, at 22 months, macaques already show classificatory behavior in both their grouping and selection of objects. This is confined, however, to only minimal two-object classes and two-object successive selections.

This picture does not progress with age. At 34 months, macaques classify in only one of the 6 possible three-object cases, but at the same time they do not show classification in one of the 6 possible two-object cases, and thus their overall rate stays at 50%, as at 22 months.

Classification appears, however, to be systematic. Fig.2 shows the distribution of the same 6 possibilities, for composing and selecting at each age, divided by class condition. As it can be seen, at both ages and in both

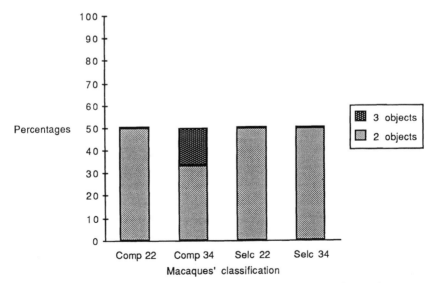

FIG. 1   Proportion of cases in which macaques exhibit classificatory behavior in Object Composing and Object Selecting at each age divided by size of compositions and selections.

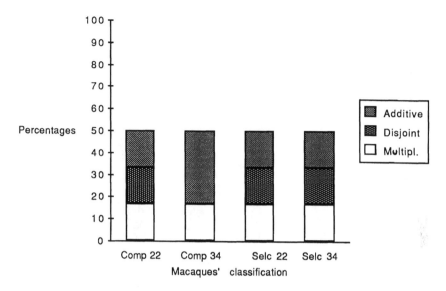

FIG. 2. Proportion of cases in which macaques exhibit classificatory behavior in Object Composing and Object Selecting at each age divided by class conditions.

composing and selecting, classification is present in all three class conditions, with the only exception of composing at 34 months, which is random in the Disjoint condition, but systematic in both cases (with two and three objects) in the Additive condition. Examination of the type of classificatory behavior shown in the different class conditions seems to confirm that macaques are indeed paying attention to properties determining class membership of objects, rather than operating on the basis of some object-saliency strategy. Fig. 3 shows the distribution of the same 6 classificatory possibilities, for composing and selecting at each age, divided into those where systematic class-consistency was manifested (labeled classification by similarity) and those where systematic class-inconsistency was manifested (labeled classification by difference).

Classification by similarity marks 9 of the 12 cases in which classification was actually manifested, while the 3 remaining cases were by difference.

Classification by similarity is not only dominant everywhere, but at the youngest age it takes place just in the two ambiguous class conditions, i.e. Additive and Multiplicative, while in the only non-ambiguous class condition, Disjoint, classification is by difference. This result runs both ways opposite to the hypothesis, advanced by several investigators to explain young children early classificatory behavior and discussed in the introduction above, that (apparent) classification might result from successive selecting identical objects on the basis of saliency. In fact, in the Multiplicative

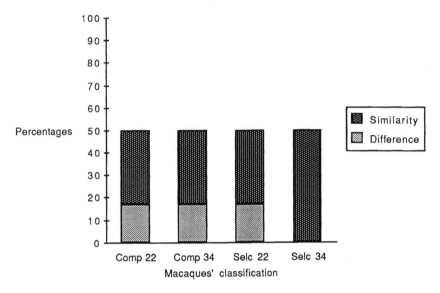

FIG. 3. Proportion of cases in which macaques exhibit classificatory behavior in Object Composing and Object Selecting at each age divided by
type (similarity or difference) of classification.

condition there are no identical objects and classifying by similarity implies
paying selective attention to either form abstracting from material or to
material abstracting from form. In the Additive condition such a strategy
would work, because there are identical objects, but they are more "con-
fused" than in the Disjoint condition, because all the 6 objects share one
property (for example, they are all of identical shape). Hence, a reasonable
prediction of the saliency strategy would be that classification by similarity
should be easier (more frequent and/or earlier to take place) in the Disjoint
condition, where the 6 objects are neatly divided into two classes that do
not resemble at all to each other. By the same token, classification by
similarity should be hardest in the Multiplicative condition. On the other
hand, we find that at 22 months macaques classify by similarity, in both
composing and selecting, in the Multiplicative and Additive, and by difference in the Disjoint condition: where identical objects are present and
are maximally distinguishable from non-identical objects, macaques systematically select and put together objects that are different.

At 34 months, all classification in Object Selecting becomes by similarity,
in all three class conditions. In Object Composing, instead, while in the
Additive and Multiplicative conditions the tendency to classify by similarity
strengthens even more, the Disjoint condition becomes random, which
might be indicative of progressive shift to classification by similarity all

over (especially, if taken together with the shift to classification by similarity in the same condition in Object Selecting). On the other hand, the only case of three-object classification, in the Additive condition, is by difference. One possible interpretation of this result (but a highly speculative one, in absence of other corroborating data) might be that there is a general tendency for classificatory activity to begin by difference and shift to similarity (an hypothesis, for example, explicitly advanced for human infants by Langer, 1980; 1986). If this were true, then one could expect that at each level of complexity within each condition, i.e. first at the level of two objects and than at the level of three objects, one would find first classification by difference and then by similarity. In this case, since three object classification begins only at 34 months in the Additive condition this should go by difference. The shift from classification by difference to classification by similarity in the two-object Disjoint condition would be explained by the same hypothesis, and this would be the only other supporting evidence for this speculation. On the other hand, one would have to assume that the "by difference" classification phase of two-object Additive and Multiplicative condition has already passed by 22 months of age; an assumption that is not contradicted by our results, but which there is no evidence for, either.

A final interesting observation is that where choice is offered to classify either by similarity in form or by similarity in material, i.e. in the Multiplicative condition, macaques consistently choose to go by material, both at 22 and at 34 months. The criterion for class membership appears to be rigid: only one of the two possible ways of classifying the same objects is chosen.

To sum up, macaques do seem to display real classificatory abilities even at their youngest age. They are, however, confined to minimal selections and groupings of two objects, and, most important, do not seem to undergo any substantial development over a whole year period. The organization of objects into classes is, furthermore, not flexible at both ages.

Fig. 4 gives a graphic summary of cebus' overall classificatory performance at each age considered. The diagram has been constructed as the corresponding one of macaques (Fig. 1): it shows in what proportion of the 6 possible cases in Object Composing (COMP) and Object Selecting (SELC) cebus manifested systematic classification at each of the three ages.

At 16 months, classification is confined to 1 out of 6 possibilities in composing. At 36 months it extends to half of the possible cases (3 out of 6) and it finally covers all cases but one (5 out of 6) at 48 months. This gradual progression is accompanied by a similar progression in the number of objects included in the systematic classes thus formed, also shown in Fig. 4. At 16 months, the only case of classification is with two objects; at 36 months all the two-object group-types show classification, but none of

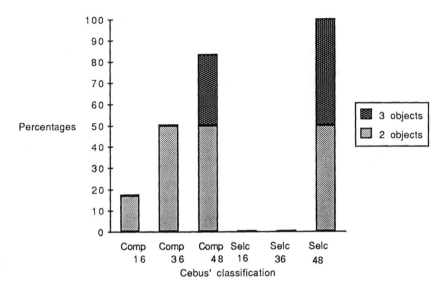

FIG. 4. Proportion of cases in which cebus exhibit classificatory behavior in Object Composing and Object Selecting at each age divided by size of compositions and selections.

the three-object ones; at 48 months all of two-object and 2 out of the 3 possible three-object group-types are classified.

Classification in Object Selecting is, instead, completely absent at both 16 and 36 months, but it reaches its maximum possible level (6 out of 6) at 48 months: at this age the sequential selection of both two and three objects takes into account the class properties of objects in all class conditions. The overall picture is that of a clear and sustained development in classificatory abilities.

As in the macaques' case, a real capacity to classify seems to be present since the earliest age considered. It is true that at 16 months classification is present in only one class condition, but it is the Multiplicative condition and it is furthermore by similarity, as shown in Fig. 5 and Fig. 6, which show the distribution of the cases of classification among the three class conditions (Fig. 5) and between the class-consistent vs. class-inconsistent type.

Thus at 16 months cebus are able to classify in the most ambiguous of the three class conditions, where there are no identical objects, systematically grouping together objects that are only similar with respect to one property.

At 36 months classification is present in all class conditions: it is by difference in the Disjoint and Multiplicative conditions and by similarity in the Additive condition. For the same reasons discussed above in the

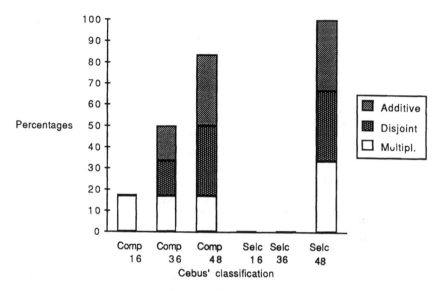

FIG. 5. Proportion of cases in which cebus exhibit classificatory behavior in Object Composing and Object Selecting at each age divided by class conditions.

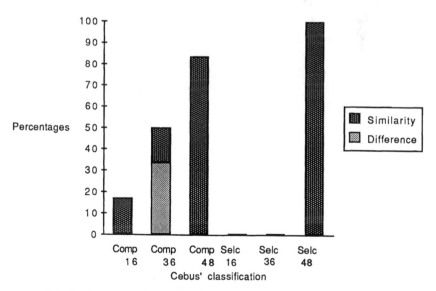

FIG. 6. Proportion of cases in which cebus exhibit classificatory behavior in Object Composing and Object Selecting at each age divided by type (similarity or difference) of classification.

macaques' case, this pattern excludes that a simple object saliency strategy is responsible for the classificatory behavior, and seems to confirm that our subjects are paying attention to the properties defining, in each class condition, the class membership of the objects they are dealing with. Finally at 48 months classification is found all over, in both composing and selecting, and, as it can be seen in Fig. 6, it is in all cases by class consistency: wherever they are grouping objects together or sequentially selecting them, cebus systematically choose objects that belong to the same class. There remains to be explained the unique presence of classification by difference at 36 months. Unfortunately, the hypothesis formulated for the macaques, of a general developmental precedence of classification by difference over classification by similarity, does not work in this case. In the Multiplicative condition, classification shifts from similarity (at 16 months) to difference (at 36 months) back again to similarity (at 48 months). In the Additive condition, the passage is from no classification (at 16 months) directly to classification by similarity (at 36 months). Only the Disjoint condition conforms to the predicted sequence. It remains true, however, that, over all, classification by difference is more confined and occurs at a comparatively earlier age.

Finally, contrary to the macaques, cebus do not show a strict preference for one of the two class properties of objects when they have a choice to classify either by similarity in form or by similarity of material. They consistently choose form over material (thus doing the opposite of what macaques do) only in two of the four cases where they classified by similarity in the Multiplicative condition. In the other two cases, they sometimes choose to follow form as the class membership criterion and some times material. This flexibility might possibly denote a more general construction of the notion of class membership: belonging to a class means to have a property as such in common, while it is totally indifferent which one is the specific property.

CONCLUSION

The analysis of the spontaneous groupings of objects produced by our subjects shows that both species possess a capacity for logical classification: they are capable of following systematically class membership criteria in a variety of different situations where these criteria differ, and are not confined to choosing physically identical objects or completely different objects, but require some abstraction of criterial properties. The two species differ, however, in the extent and flexibility of application of this basic capacity and in its developmental progression. Macaques appear to be limited to the construction of only minimal class-coherent collections and do not alternate possible classification criteria. Furthermore, they do not seem to develop and expand their classification capacity along any of these dimensions. Cebus, on the other hand, show a consistent and prolonged development, thereby beginning with minimal two-object collections at 16 months they become capable of classifying larger collections at 48 months according to consistent criteria of similarity, and are capable, furthermore, of flexibly alternating dimensions determining class membership.

How do our species compare with human infants' classification development? Langer (1980, 1986) analyzed the spontaneous groupings and selections of two sets of objects related by the same class conditions (Additive, Disjoint, Multiplicative) we offered. The first instance of children's classification by similarity occurs at 12 months and by 18 months they are capable of classifying by similarity in all three class conditions, their progression being from Additive to Disjoint to Multiplicative. A more quantitative comparison reveals that at this last age children classify over all in only half of the possible cases offered: 2 out of 6 for composing, and 4 out of 6 for selecting. Before 12 months no classification by similarity occurs. Taking together all possible cases of classification offered between 6 and 10 months, children show systematic classification behavior in 5 out of 18 possibilities, while in the remaining 13 their groupings and selections are random with respect to class membership. Quite interestingly, all these instances of classification are by difference, and, furthermore, 4 out of 5 occur at 6 months of age. At this age, classification by difference is present in all class condition.

At this large-scale level, there seems to be a developmental progression from classification by difference to classification by similarity with an intermediate period between these two where no classification occurs. If, indeed, this is the correct interpretation, then the turning point of this progression occurs at 12 months, and, in fact, at this age classification begins to be by similarity in the Additive condition, it is random in the Disjoint condition and still by difference in the Multiplicative condition.

This trend, however, is not so linear and clear at a finer scale: for example, in the Multiplicative condition, classification is by difference at 6 months, then becomes random at 8 and 10 months and then it is again by difference at 12 months, before beginning to be by similarity at 15 months; in the Disjoint condition it is by difference at 6 months, random at 8, by difference again at 10 and then random again at 12. A weaker and more conservative statement might be that, however, wherever classification by difference is present it always precedes classification by similarity.

When classification by similarity begins, there seems to be also some evidence that classification in selecting precedes classification in composing, a finding also reported in former studies of early classification. At 15 months, only 1 out of 6 possibilities of classifying in Object Composing is realized, but at the same age 5 out of 6 are realized in Object Selecting. At 18 months the figures are 2 out of 6 in composing and 4 out of 6 in selecting and at 21 months 3 out of 6 in composing and 4 out of 6 in selecting.

As we saw above, the hypothesis that classification by difference precedes classification by similarity is not incompatible with the macaques' data. It is, instead, contradicted in one case by the cebus data, though, at a more general level, even in this species classification by difference is present at 36 months and disappears entirely at 48 months. Thus at least a tendency for classification by difference, if present, to precede classification by similarity (within each condition and level) seems to be common to all species.

Overall, macaques classificatory abilities seem to match, in type and quality, that of children of 12–15 months: classification is present in all class conditions (though not performed in all possibilities) with a mixture of "by similarity" and "by difference" criteria; it is, furthermore, present by similarity in all three class conditions only in selecting (at 34 months), as in 15 month-old children. As we saw, however, this is not the result or the start of an on-going developmental progression; more specifically, the extent of collections thus classified remains at all times minimal, while 15 month-old children already classify within classes of 4 objects each.

Cebus, on the other hand, reach this same level of classificatory abilities at 36 months, but they show a definite developmental progression from 16 months and continue to develop afterwards. At 48 months, their level is, in some respects, even higher than 24 month-old children: cebus classify in all but one of the possibilities allowed in both composing and selecting, always by similarity, while children are still random in a few of them. Thus, from the point of view of *single* class construction, i.e. of selecting and grouping objects in one class, cebus appear as or even more advanced than 24 month-old children. We underlined the term "single" because if we take classification of children at 24 months as a term of comparison, then an important difference must be made. At this age, in fact, children compose a substantial amount of *two* class coherent groupings, and, most important, they construct a portion of them simultaneously, i.e., they construct two class coherent groupings of objects at the same time (see, also, Sugarman, 1983). This is not found in our subjects: construction of two separate sets of objects either in spatial or in temporal proximity (even independently of their class properties) is extremely rare in both macaques and cebus, with the exception of cebus at 48 months. The relevant data are reported in Table 3 of the following chapter, where relatedness between compositions is analyzed. At 48 months, there is a certain amount of constructions resulting into two sets of objects: 60 of them can be considered for classification (the remaining include fragments of objects) and only in 23 occasions the two sets are class consistent, but in no case there has been simultaneous construction of the two classes. Of the 28 cases where, instead, there has been simultaneous construction of two sets, in 1 case only the two resulting sets are both class consistent.

In cebus, therefore, the capacity of dealing with class grouping grows substantially, but within the limit of single-class organization. As it will be seen in the next chapter, this pattern, whereby a logical capacity grows in frequency and extent of application but not in structural complexity, is not confined to classification.

14 Logical Operations

Patrizia Potí and Francesco Antinucci
Istituto di Psicologia, CNR
Rome, Italy

*When Ulysses had left the land of the Cyclops, after blinding Poly-
phemus, the poor old giant used to sit every morning near the entrance
to his cave with a heap of pebbles and pick up one for every ewe
that he let pass. In the evening when the ewes returned, he would
drop one pebble for every ewe that he admitted to the cave. In this
way, by exhausting the stock of pebbles that he had picked up in the
morning he ensured that all his flock had returned. . . . By long
practice in handling the abstract symbols 1, 2, 3 . . . we are liable to
forget that they are only a shorthand way of describing the result of
an operation, viz., that of matching the items of an aggregate with
those of some set of standard collections that are presumed to be
known.*

—Jagijit Singh, *Great Ideas of Modern Mathematics*, Dover, N.Y., 1959.

INTRODUCTION

Long before acquiring any language capacity, human children appear to
develop and master fundamental logical abilities, that are constructed from
(and manifested in) their sensorimotor action on their environment. When
appropriately investigated, sensorimotor precursors of logico-mathematical
structures, such as class-inclusion, identity, equivalence, correspondence,
inversion, reciprocity, etc. have been found to characterize the structure
of the child's spontaneous interactions with manipulable objects beginning
from as early as 6 months (Langer, 1980, 1986; Forman, 1982; Sinclair et
al., 1982).

Logical knowledge arises from structured sequences of actions, i.e., sequences that become progressively regulated by the product of the actions themselves. In the traditional Piagetian language this process is known as "coordinations of actions" and it is these coordinations of actions that, once reflected upon and internalized in representation, give rise to true logical structures:

> These coordinations result in the construction of a sort of logic of schemes prior to language and thought. At the heart of it one can already find the main type of connections-relationships of order, embedding of schemes, correspondences, intersections, some form of transitivity, associativity of monoids, etc.—which are the main ingredients of future operatory structures. (Piaget, 1976, p. 351, translation revised)

Tracing the foundation of logical capacities, therefore, amounts to tracing the extent and ways through which action sequences become structured.

The building blocks of logical operations are the actions of Composing and Decomposing: i.e., in their most general form, putting separate objects together and taking them apart. Composing and Decomposing generate the basic objects of logical operating: sets and elements as constants.

Sequences of Composing and Decomposing that are structured, that is, applied to and regulated by their successive products, generate all other logical operations.

We can group them into two general classes. Those that give rise to identity and equivalence, and those that give rise to ordered nonequivalence or quantitative relations. The first class includes biunivocal correspondence, exchange operations and negation. The second class includes addition, subtraction, counivocal correspondence, multiplication and division. Biunivocal correspondence constructs equivalence between sets and is the essential component of the number concept. Exchange operations construct the conservation of the invariance of a set through successive intensional and extensional transformations. Negation operations also construct invariance, through inversion or reciprocity, i.e., by generating the fundamental structure of reversibility.

Addition and subtraction generate sets regularly and systematically variable in extent, thus establishing the basis for quantitative relations. Counivocal correspondences, that is regulated repeated correspondences of one-to-many or many-to-one, between two sets is the prerequisite operation for multiplying or dividing two sets by each other. Finally, multiplication and division form the basis of distributional relations. In what follows we will illustrate each of these operations and their relationships, the mode and extent of their construction by nonhuman primates, and the way they compare to those of the human infant.

DATA-ANALYSIS

The general criterion followed to identify logical operations and their precursor components involves a three-step procedure: 1) Define the action sequence corresponding to the full-fledged operation; 2) Subtract components and/or weaken conditions of occurrence; 3) Use the product thus obtained to check actual results.

For example, the full-fledged sequence defining an additional operation would be to construct a series of sets that increase regularly in numerosity (2 objects, 3 objects, 4 objects, etc), in spatio-temporal contiguity (Step 1). A weaker condition would be to accept also a minimal component of that sequence, i.e., a series of two sets one of 2 and one of 3 objects (Step 2). A still weaker condition would require to construct a set of 2 objects and then transform this same set into a 3-object set by adding one object: in this way it is not necessary to deal with two sets of objects at a time (Step 2). Check the results and score only those transformations of a set numerosity that are characterized by regular increments (Step 3). Let's now examine each operation in turn, beginning with the basic ones of Composition and Decomposition.

Composing

Composing and decomposing can be applied to both elements and sets. Composing elements generates sets; composing sets generates second-order sets, i.e. sets of sets.

A set is composed when two or more objects are put together. In order to score such a composition at least one of the objects has to be actively manipulated and the objects must be placed either in contact or at a maximum distance of 10 cm from each other. Since operations are related sequences of compositions it is essential to state criteria for relatedness between sets. The minimal such criterion is that two compositions are considered to be related if they are composed one after the other within 5 seconds. Of course, in this case we don't have two sets simultaneously present, but rather one set is transformed into another. Compositions satisfying this criterion will be labeled "temporally sequential" as opposed to "temporally isolated" ones. In all other cases of relatedness (at least) two real sets are instead present. The two sets can be constructed at the same time, in which case they will be labeled "temporally overlapping", or they can be related spatially. For this to occur, the two sets must be clearly partitioned, but, at the same time, no more distant than 20cm from each other. Obviously, second order operations, that is, operations whose elements are sets, require the simultaneous presence of at least two sets. The same criteria defining relatedness between sets apply when sets are generated by Decomposition, i.e. taking away elements from already formed sets, instead of Composition. Decomposition, however, can also be applied (though with quite different results) to single objects: when an object is broken into pieces. On the one hand, this operation changes the number of objects available, and, at the same time, the fragments so obtained lack controlled shape properties.

Once generated, however, fragments can be composed into sets, which can be operated upon as any other set. For these reasons, compositions with fragments have been scored and used in the analysis of logical operations where shape properties of component elements are not relevant. They have not been used to assess within-set correspondences (see below) and, most important, they have not been used in the analysis of classificatory structures, reported in the preceding chapter (this difference is responsible for the difference in figures regarding number of compositions taken into account between this and the preceding chapter).

Ordered nonequivalence operations

Addition and Subtraction

The conditions defining addition have already been given above, in the example. The same conditions define a subtraction operation, if the numerosity of the set is decreasing instead of increasing. Thus, in its most advanced form, the operation of addition is manifested in the construction of a series of related sets, each one unit more numerous than its predecessor, and viceversa for subtraction, as pictured in Figure 1.

As said above, a more primitive construction would be the successive transformation of a single set by adding (or subtracting) one unit at a time. Of course, the successive sets have to be temporally related as defined above.

The most important parameter in these constructions is how many times the operation is reiterated. Reiteration, in fact, produces regular seriation of quantities: 2, 3, 4, etc. (or 4, 3, 2). Hence, we will speak of one-step series (2, 3 or 4, 5 or 4, 3, etc.), two-step series (2, 3, 4 or 5, 4, 3, etc.), three-step series, etc. Addition and subtraction can be combined and produce bidirectional series, i.e., ascending-descending or descending-ascending series. Criterial is always the presence of regular one-step increments or decrements. Minimal such series are 2,3,2 or 3,2,3. In these series the second operation inverts the first and thus reestablishes the original state. In this context, therefore, the second operation also represents a negation operation, that of inversion. In general, appropriate sequences of addition/subtraction (or viceversa) (which in themselves are regulated sequences of Compositions and Decompositions) can give origin to negation operations.

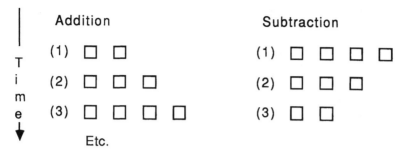

FIG. 1. Diagram of the action-construction of addition and subtraction. The identity of the elements is irrelevant.

In this case too, length of the series is an important parameter, but different structures must be carefully distinguished. In a series like 2,3,2,3, inversion is iterated twice (first in going back from 3 to 2 and then in going back from 2 to 3); in 4,3,4,3,4, likewise, we have three successive inversions. These will be called *one-step* inversions. Quite different is a series like 2,3,4,3,2, where the whole 2–3–4 series is completely inverted. Since in this case a two-step series is inverted we will call this operation a *two-step* inversion. Thus series like 2,3,2,3 (or 4,3,4,3,4) will have two (or three) successive one-step inversions, while series like 2,3,4,3,2 will have one two-step inversion. In the first case we have the (repeated) inversion of an operation, in the second case the inversion of a whole structured sequence, of a series.

Multiplication and Division.

In its simplest form, multiplying one set by another set means reiterating the first set a number of times equal to the second set (2×3 means reiterating 2 three times). The equivalent in action of this operation consists in iteratively combining each element of a set in the same way and with the same number of elements of another set, as diagrammed in Figure 2.

This operation consists therefore of repeated compositions of a special character, where one object is *simultaneously* related to a number of (at least 2) other objects. This type of composition can be considered the equivalent in action of counivocal correspondence, and, more specifically, of one-to-many correspondence. Thus, multiplication consists of repeated one-to-many correspondences of the same form.

One typical case of one-to-many correspondence is the so-called "bridge construction", where one object is put on top of two others to bridge them. A minimal multiplication (i.e., 2×2) would be the reiteration of a bridge construction (see Fig. 3).

2 x 3

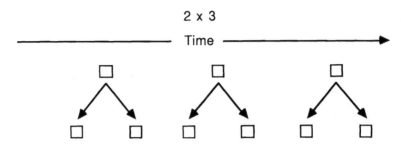

FIG. 2. Diagram of the action-construction of multiplication. A 2 × 3 multiplication is shown. A set of 2 is replicated a number of times equal to the number of elements (3) of another set. This relation is indicated by the reiteration of a one-to-many correspondence, symbolized by the arrows, between each element of the second set and each set of 2.

FIG. 3. An example of a minimal multiplication (2 × 2). A "bridge-construction," representing a 1-to-2 counivocal correspondence, is reiterated twice.

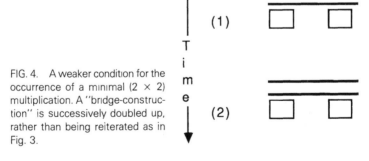

FIG. 4. A weaker condition for the occurrence of a minimal (2 × 2) multiplication. A "bridge-construction" is successively doubled up, rather than being reiterated as in Fig. 3.

A weaker condition of occurrence would be the reiteration of this construction **on the same set,** i.e., putting an object on two others and then putting another object on top of the same construction (see Fig. 4).

As in other cases, in this instance we deal with only one set at a time. A still weaker condition would be **replacing** the bridge-element with another bridge-element. In this case, in fact, not even a trace of the first one-to-many correspondence would be left. Finally the occurrence of a single one-to-many correspondence can be considered as the minimal precursor of multiplication.

6:2

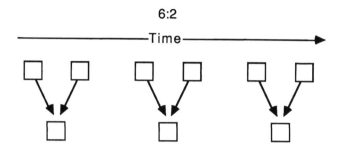

FIG. 5. Diagram of the action-construction of division. A 6:2 division is shown. Successive equal sets of 2 are taken from a set of 6. This construction is indicated by the reiteration of a many-to-one correspondence, symbolized by the arrows, that singles out each of the subsets of 2.

Division is the inverse operation of multiplication: it entails partitioning one set into a number of equal subsets. Its manifestation in action can take, however, two different forms. One of them is the exact inversion of the sequence of multiplication: a given number (at least two) of elements of a set are simultaneously combined with one element of another set and this operation is repeated in the same way and with the same number of elements (see Fig. 5).

Thus, division too consists of repeated counivocal correspondences of the same kind, but they are *many-to-one* correspondences rather than one-to-many as in multiplication. A typical case of many-to-one correspondence is, for example, putting simultaneously two (or more) sticks into a cup. A minimal division (4:2) would be the reiteration of this correspondence. The same weaker conditions seen for multiplication apply to this construction.

Division, however, can be constructed in action through a second and different procedure. To distinguish between them, we shall call the construction just described "division-by", and the one we are about to illustrate "division-into". For the sake of clarity, we will illustrate it through a concrete example. A stick is put into a cup, a second stick in a second cup, and a third stick into a third cup. Then the same sequence is repeated by taking other sticks but putting them into the same three cups, and so on. In this way, we will have *divided* the set of sticks *into* three, the number of cups.

The results of both these procedures is the same: in both cases we partition a set into subsets of the same numerosity, but through two different routes. In the first case, the numerosity of the subsets is given and the operation results into the number of subsets that are formed. In the second case, the number of subsets is given and the operation gives as a result the numerosity of each subset. The relation between these two proc-

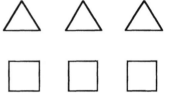

FIG. 6.   Example of biunivocal correspondence at the action level: two sets of equal size are matched element to element through spatial symmetry.

esses constitutes the basis for the exchange relation existing between divisor and quotient in a division operation ($6:2 = 3$ and $6:3 = 2$).

This second division operation consists then of *sequences of sequences* of one-to-one correspondences (see below under *Biunivocal correspondence*) of a specific and constrained kind. Each sequence has to be performed onto the same correspondent set of elements, so that the final (static) result is a set of many-to-one correspondences. It is obvious that in scoring for this operation components cannot be used, since they are indistinguishable from simple one-to-one correspondences: at least two (minimal) sequences with the required characteristics must occur.

Equivalence operations

Biunivocal correspondence

The fundamental operation that gives rise to (discrete) quantitative equivalence is biunivocal correspondence. Two sets are equivalent if they can be exhaustively matched element to element. On the action level, evidence of this operation comes mainly through some kind of spatial regularity in either the result or the process (or both) of matching the elements of two sets (see Fig. 6).

Important quantitative parameters are the numerosity of the sets matched and, especially, whether the matching is exhaustive. Two construction processes of biunivocal correspondence can be distinguished. In the most advanced case, both the construction of the two sets and the biunivocal matching proceed at the same time, in an interweaved fashion. Thus, the two sets are in both spatial and temporal relation. In the less advanced case, first a set is constructed and then a second set, by matching the elements of the first. The two sets are in spatial relation only. With reference to Fig. 6, the two processes can be represented as follows (where numbers denote order of placement):

	A			B	
1	3	5	1	2	3
2	4	6	4	5	6

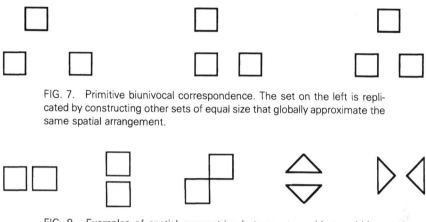

FIG. 7. Primitive biunivocal correspondence. The set on the left is replicated by constructing other sets of equal size that globally approximate the
same spatial arrangement.

FIG. 8. Examples of spatial symmetries between two objects within a
set.

A more primitive case can be considered the one in which the elements
of the second set are not exactly put in correspondence one by one with
those of the first set, but replicate, with higher or lower approximation,
the spatial arrangement of the first set. An example is shown in Fig. 7.

In general, all cases of symmetries between two sets of objects (or two
subsets of a set) can be considered precursors of biunivocal correspondence: symmetries are in fact generated by (partial) correspondences between parts of a (complex) "object". For this reason, one can consider as
the most primitive precursor of biunivocal correspondence the construction
of spatial symmetries between two (or more) objects **within** a single set,
such as precisely aligning two objects on a plane, or stacking them in a
tower-like construction, or arranging them in a mirror relation, as depicted
below. In all such cases, in fact, the symmetries result from global spatial
correspondences between parts of the objects as diagrammed in Fig. 8.

Exchange operations

Exchange operations establish the quantitative invariance of a set under
transformations affecting its constituent elements. They are therefore an
essential component of the number concept. More specifically, exchange
operations construct equivalence between two consecutive sets where one
(or more) element has been varied and the other(s) have been kept constant. Transformations of a set resulting into a quantitatively equivalent
set can be of two kinds: one can vary the spatial placement of its constituent
elements relatively to each other (their *order*) or one can vary the physical
identity of its constituent elements. Accordingly, we have a *commutativity*
operation, where one element is taken out of a set and then recomposed

in a different order relation, and a *substitution* operation, where an element of a set is taken out and then replaced with a physically different element. Diagrammatically,

	Commutativity			**Substitution**	
		C		B	
(1)	A	B	A		C
				B	
(2)	A	C B	A		D

Commutativity is an essential component of conservation: quantity is conserved under transformations of shape, form, density, etc. Substitution, on the other hand, is essential in constructing the notion of an abstract quantitative unit, which is independent of the nature of the individual object that embodies it, and as such represents a more complex operation. As a precursor of these transformations that preserve identity when the constituent elements of a set are affected, it is useful to consider a more simple operation than commutativity and substitution, representing a transformation that preserves identity because it is "empty". This occurs when an element is taken out of a set and then the same element is put back into it in the same position, thus leaving unchanged both the order and the constituency of the set. We will call this operation *replacement.*

All three exchange operations also help to construct the relationship between *constant* and *variable.*

In this connection, **sequences** of commutativities, substitutions and replacements linking successive transformations of a set can feature important parameters. One of this is whether (or to what extent) **all** the elements of a set are operated upon one after the other. The second is whether successive operations reconstitute the original set, i.e., whether they are also negation operations.

The simplest case is that of the application of two consecutive commutativities on a two-element set. First the order of A and B is changed by operating on A:

(1)	A	B	
(2)		B	A

Then there are two possibilities. The order of A and B is again changed by operating on A:

(3a)	A	B

In this case the second commutativity reestablishes the original order by inverting the result of the first. In this sequence it is, thus, a negation operation, and, specifically, an inversion. In the second possibility, the order of A and B is again changed, but this time operating on B:

(3b)   A   B

In this case also, the order of the original set is reestablished, not, however, by inverting the first operation, but rather by **compensating** for it. The negative operation is, in this case, a *reciprocal* operation and not an inversion. Sets with more than two elements can give rise to sequences of commutativities with more complex structures of inversion and reciprocity.

Two structures can be distinguished in sequences of consecutive substitutions. In one case, one and the same element is kept constant through a series of substitutions. In a two-element set this takes the form of

(1)   A   B
(2)   A   C
(3)   A   D
(4)   etc.

This structure has the form of a function $Cx = y$, where C is the element kept constant, x the variable, and y the resulting set. It establishes the functional equivalence of all the objects that can be substituted for x, and hence the abstract class of units. For this reason, the most important parameter of this type of sequences is their exhaustiveness, i.e. to what extent the substitutions in the x-role tend to be applied to all the objects present.

The second type are sequences where all the elements at disposal are systematically permuted into a set. For example, in a two-element set ranging over a universe of three elements, a sequence like

(1)   A   B
(2)   A   C
(3)   B   C

Here the function is of the form $xy = C$. Any pair of objects can fill the slots for the (constant) construction, and they can be interchanged with each other in their roles. Such sequences can also be interested by negation. There can be cases of inversion, such as (continuing the above sequence)

(4)   A   C

in which operation (3) is inverted and the set in (2) is reestablished. Or more complex cases, such as

(5)   A   B

where the **sequence** of operations (4) and (5) inverts the **sequence** (2) and (3). As done before, we shall call the first type a *one-step* inversion, and the second a *two-step* inversion.

This second type of sequences of substitutions also allow negation by reciprocity, such as

(1)	A	B
(2)	A	C
(3)	B	C
(4)	B	A

where the original constituency of the set (A B) is reestablished, not, however, by inverting the result of operations, but by compensating for it.

Thus, as for other operations, important features to be considered are the length of the sequences (in terms of number of iterated operations) and, especially, their internal structure.

Negation

Negation is the fundamental operation that constructs identity and it does so by reversing the effects of transformation, and thus allowing the reconstruction of the departing state. Negation operations are therefore the basis of reversibility. There are two such operations: inversion, that reestablishes the departing state by inverting the precedent operation(s), and reciprocity, that does so by compensating the effects of the preceding operation(s). Both these operations and their possible structures have been illustrated above in connection with the characteristics that sequences of other operations can show (sequences of additions and subtractions, sequences of commutativities, sequences of substitutions).

It must be noticed, in fact, that negation operations are such not by themselves, but only by their link to a preceding direct operation. Thus, they can only be found in the context of a sequence of operations: in the series 2,3,2 the subtraction operation is an inversion operation by virtue of its sequential context, because it is immediately preceded by an addition.

RESULTS

Compositions

Number

Table 1a reports compositions generated by each species at each age point according to their size (M = macaques; C = cebus; number indicates age in months).

TABLE 1a.
Combinativity Operations: Set-Size

Age in Months	Two-Object Sets		Three-Object Sets		Four-or-More-Object Sets		Total	
	N	%	N	%	N	%	N	%
M22	452	82.3	48	8.7	49	9	549	100
M34	323	74.8	72	16.6	37	8.6	432	100
C16	422	81.3	62	12	35	6.7	519	100
C36	721	70.5	191	18.7	110	10.8	1022	100
C48	1515	59.8	607	24	410	16.2	2532	100

TABLE 1b.
Percent Distribution of Set-Size in Children Compositions

Age in Months	Two Object Sets %	Three Object Sets %	Four-or-More Object Sets %	Total %
H 8	82	15	3	100
H12	75	18	7	100
H15	63	22	15	100
H18	54	15	31	100

Total number of composition generated decreases with age in macaques, while it strongly increases (roughly doubling at each age point) in cebus; at both 22 and 34 months macaques generate a number of compositions comparable to that of 16-month-old cebus.

Size

In both species minimal, two-object sets represent the vast majority of all sets composed. Their proportion decreases with age in favour of an increase in the proportion of larger sets. This trend, however, differs in the two species: in macaques the increase is confined to sets of three objects, while sets of four or more objects remain stable; in cebus, both three-object and four-or-more object sets increase regularly with age. Statically, there is a good correspondence in set size between 22-month-old macaques and 16-month-old cebus, on the one side, and 34-month-old macaques and 36-month-old cebus, on the other side. Cebus at 48 months, show a substantial progress in the same direction.

To compare set size of compositions generated by nonhuman primates to that of human infants, Table 1b shows the proportion of compositions of different sizes generated by children between 8 and 18 months of age (H = human child; number indicates age in months; data drawn from Langer, 1980, 1986).

Since from 12 months on children received both sets of four objects and sets of eight objects to manipulate, percentages reported for these ages represent averages over the two conditions). Cebus at their oldest age (48 months) match children at 15 months. Cebus at 36 and macaques at 34 months match children at 12 months. The fit is instead less good for cebus at 16 and macaques at 22 with children at 8 months: the proportion of two-object compositions is the same, but macaques and cebus have a higher proportion of four-or-more object sets with respect to children while these have a higher proportion of three-object sets. These differences, however, can be easily explained by the fact that 8-month-old children had only four objects to manipulate, and, therefore, for them four-object sets were the largest possible and implied putting together all the objects presented. In fact, the size of sets constructed in the 8-object condition is consistently larger on the average than the size of sets constructed in the 4-object condition at all ages where this comparison can be made (see Langer, 1980, 1986). The same phenomenon can be seen in our subjects, where the presence of fragments, that increase the number of "elements" available, increases the proportion of larger compositions (compare the data on four-or-more-object sets in Table 1a, which include compositions with fragments, to the same data in Table 1 and Table 7 of the preceding chapter, where such compositions are excluded). Overall, a similar pattern of development (obviously, with very different age and temporal correlates) characterizes the three species on this dimension of composing.

Temporal features

Table 2a shows temporal features of composing, subdividing compositions into those composed in isolation, consecutively and in temporal overlap.

As defined above, the minimal requirement for the occurrence of an operation is that two sets be related at least in time (i.e., one set transformed into another in some regulated fashion), even if in this case only one set at a time is present. Therefore, consecutive compositions, but not isolated ones, generate single sets susceptible of being related by operations. Temporally overlapping compositions generate, on the other hand, at least two contemporaneous sets, which is the minimal prerequisite for *second-order* operations to occur.

As shown in Table 2a cebus monkeys did not show any developmental difference between 16 and 36 months in the proportion of isolate vs. coordinate compositions or in the proportion of sequential vs. overlapping compositions. Only at 48 months an increase did occur in both sequential and overlapping composing, accompanied by a strong decrease of the proportion of isolate compositions. The macaques' pattern is different. Isolate compositions decreased with age and correspondingly sequential compositions increased, but overlapping compositions strongly decreased, though starting at a quite high level.

Table 2b reports temporal features of children composing (data taken from Langer, 1980, 1986 and treated as in Table 1b). Contrary to set size, the comparison reveals a different developmental pattern. Between 22 and 34 months macaques increase the proportion of sequential composition reaching the level of 10 month-old children, but at the same time the proportion of overlapping composition follows an inverse course decreasing from the level of 10-month-old children to that of the 8 month-old. Cebus at 48 months are more advanced than even 12-month-old children in their

TABLE 2a.
Combinativity Operations: Temporal Features

| Age in Months | Temporally Isolated Sets | | Temporally Related Sets | | | | Total | |
| | | | Sequential | | Overlapping | | | |
	N	%	N	%	N	%	N	%
M22	143	26	384	70	22	4	549	100
M34	96	22.2	330	76.4	6	1.4	432	100
C16	146	28.1	371	71.5	2	0.4	519	100
C36	319	31.2	701	68.6	2	0.2	1022	100
C48	404	16	2073	81.8	55	2.2	2532	100

TABLE 2b.
Percent Distribution of Temporal Features of
Compositions in Children

Age in Months	Isolated Sets %	Sequential Sets %	Overlapping Sets %	Total %
H 6	34.5	64.5	1	100
H 8	37	62	1	100
H10	18	79	3	100
H12	18.5	74	7.5	100
H15	13	76	11	100
H18	6	79	15	100

production of sequential composition and low proportion of isolates, but their proportion of overlapping composition is very far from that of 12 month-old children (7.5% vs. 2.2%) and does not even reach the level of 10-month-old. Overlapping composing lags behind their general level of composing, in cebus, and strongly decreases with age, in macaques.

These data seem to indicate that while the potential for operationally interrelating successively single sets grows in both species regularly as in children (and quite strongly in cebus), that for operationally interrelating two actual sets is minimal: it grows very little in cebus, as compared to children, and diminishes in macaques. It is therefore only the potential for first order operations (relating single sets) that grows but not that for second order ones (relating at least two contemporaneous sets). In fact, in children the proportion of sequential compositions does not change substantially between 10 and 18 months; what does undergo a large increase is the proportion of overlapping compositions, growing in this period from 3% to 15%. It should be noticed that in the most favorable case (cebus at 48 months) only 56 sets out of a total of over 2,500 were constructed together.

Second order

Two sets, however, can be related *a posteriori* rather than in their construction process, i.e. they can be constructed one after the other (even in temporal isolation from each other) but be related in their spatial disposition. For this to occur, they have at least to be in spatial contiguity according to the criteria specified above.

Therefore, as a check on the potentiality of second-order composing, all cases where at least two sets are simultaneously present and spatially contiguous were tabulated. They are reported, together with all cases of at least two sets constructed in temporal overlap, in Table 3, in order to give the total number of occasions where two actual sets could be related.

TABLE 3.
Prerequisites to Second-Order Composing

Age in Months	Temporal Overlap	Spatial Proximity	Temporal Overlap & Spatial Prox.	Total	Operationally Related
M22	5	2	6	13	2
M34	—	18	3	21	2
C16	1	1	—	2	—
C36	1	5	—	6	1
C48	14	101	14	129	21

Finally, the last column of Table 3 reports in how many of these occasions two sets were indeed operationally related.

As it can be seen, for both species at all ages, except cebus at 48 months, figures are extremely low. Macaques show a minimal presence of potential second order composing, and an insignificant realization of this potential (2 cases at both 22 and 34 months). In 16-month and 36-month-old cebus this potential is instead practically absent. 48-month-old cebus, on the other hand, do show a certain amount of cases where two sets were constructed in proximity and/or at the same time, but the number of cases where they were in fact operationally related is again very small (21). These cases, which exhaust all instances of second-order composing, will all be examined when dealing with the specific operations involved.

It is interesting to notice that in both macaques and cebus at their oldest age, composing more that one set in spatial proximity only is much more frequent than composing more than one set in temporal overlap. This seems to indicate that whatever development of preconditions for second order composing there might be, it comes from sequential composing of single sets, rather than from the contemporaneous construction of two sets, as it happens instead in children's development (see Langer, 1980).

Ordered non-equivalence operations

Addition and Subtraction

Full-scale addition and subtraction (i.e., the construction of contemporaneous sets precisely increasing or decreasing of one unit) was found in minimal one-step series and in extremely rare instances, except for cebus at 48 months. Macaques at 22 constructed twice a minimal 2,3 series. At 34 they constructed twice a 3,4 series. No such series were constructed by 16-month-old cebus and one 3,2 series by 36-month-old cebus. As it can be seen, these minimal series represent also all cases of second order composing in subjects at these ages (see Table 3), i.e., all cases where two fully constructed sets are precisely related by an operation.

Cebus at 48 months constructed, instead, 20 such series: ten 2,3 series, nine 3,2 series and one 5,4 series. Though at this age their number is much higher, these series remain structurally minimal and do not increase in complexity: only one increasing or decreasing step between two sets of the minimal possible extent (2 and 3 objects). Also for cebus at 48 months, these series nearly exhaust all second order composing: 20 out of 21 instances. The remaining instance is one (and the only) case of biunivocal correspondence, which will be examined below. Children start constructing second order series at 12 months with sets of 2 and 3 objects, and at 15 months relate sets of 3 and 4 objects. Contrary to those constructed by the nonhuman primates, children series tend, furthermore, to show features of spatial correspondence between the two sets.

First-order series, where one set is successively transformed by precise 1 element increases or decreases, are, instead, well established in all subjects. Table 4 reports all occurrences of both increasing and decreasing series.

As it can be seen, a few of them (two-step series) are structured beyond the minimal level, extending to three compositions, related by two successive addition or subtraction operations. One-step series are substantially stable and present at about the same level in macaques at 22 and 34 months and cebus at 16 and 36 months: about a fifth of all compositions constructed are related by an addition or subtraction operation. Again there is a large quantitative increase for cebus at 48 months: the proportion of compositions thus related represents 42% of all compositions. Also at this age and in this species there is a fair number of the structurally more complex two-step series. At this age there were even two instances of three-step series, i.e., a series of 4 compositions related by regular 1 element increments: 6,7,8,9 and 8,9,10,11. The proportion of these more extended series is, however, extremely low in both species, though it does increase with age.

Additions and subtractions can follow each other and generate bidirectional series. These inverted series can be of various degrees of structural complexity (one-step inversions, e.g., 2,3,2; two-step inversions, 2,3,4,3,2). Table 5 reports all occurrences of such inverted series, and, as a measure of how systematically negation occurs in these structures, the percentage of inverse operations on the total number of operations performed.

Two results stand out. On the one hand, negation by inversion of the preceding operation is fairly systematic in one-step series. In cebus, over a third of additions and subtractions performed represent inversions of the preceding operation; there is no developmental progression between 16 and 48 months, but instead a small drop at 36 months. Macaques start out at 22 months at a level similar to cebus, but there is a substantial drop at

TABLE 4.
First-Order Series

Age in Months	One-Step Series				Two-Step Series			
	Increasing N	Decreasing N	Total N	% on Comp.s	Increasing N	Decreasing N	Total N	% on Comp.s
M22	30	25	55	20	1	2	3	1.6
M34	23	19	42	19.4	5	—	5	3.5
C16	27	41	68	26.2	2	—	2	1 1
C36	64	84	148	19.2	2	3	5	1.5
C48	263	268	531	42.1	14	16	30	3.6

TABLE 5.
Inversions on First Order Series: Bidirectional Series

Age in Months	One-Step Inversions N	Total One-Step Series N	% of Inversions on Total One-Step Series	Two-Step Inversions N	Total Two-Step Series N	% of Inversions on Total Two-Step Series
M22	19	55	34.5	—	3	—
M34	8	42	19	—	5	—
C16	25	68	36.7	—	2	—
C36	39	148	26.3	—	5	—
C48	207	531	38.9	1	30	3.3

34 months. It seems that in these structures the inverse operation is present since the beginning of the ages considered and does not under go any structural or quantitative development; on the contrary, it tends to drop out in macaques.

On the other hand, the structurally more complex two-step inversions, where negation invests a sequence of preceding operations rather than a single operation, **never** occurred, but in one exceptional case performed by cebus at 48 months. This notable sequence was: 4,3,2,3,4.

It can be clearly shown that the nonoccurrence of two-step inversions has to do with the *structural complexity* of the series involved and not with their length. In fact, one-step inversions can and do occur in repeated non-interrupted sequences, as reported in Table 6, but they always invert the result of the immediately preceding operation and never span over two (or more) direct operations, as would be required in structurally more complex inversions.

Table 6 gives a quantitative appraisal of this phenomenon, by reporting how many of the total number of inversions occur in isolation (1 item) vs. longer non-interrupted sequences (2 items, 3-or-more-items), and their proportion over the total number of inversions performed. Examples of long sequences of inversions performed are, for example, C 36: 2,3,2,3,2; C 48: 4,3,4,3,4. Being structurally one-step inversions simply repeated, these sequences are present since the beginning and as simple one-step inversions do not undergo any developmental change. Only cebus at 48 months tend to iterate inversions in longer and longer sequences.

The level reached by macaques and 16- and 36-month-old cebus approaches that of 12-month-old children. In children, at this age, one-step series are common and are complemented by equally often occurring one-step inversions. Two-step series occur, instead, rarely. Repeated sequences of one-step inversions are frequent and from 10 months on might involve 3 or more repetitions. Two-step series become frequent at 15–18 months, when also two-step inversions make their appearance, and this corresponds to the level reached by 48-month-old cebus.

Multiplication and Division

No multiplication or division operation was ever performed by our subjects: counivocal (or biunivocal) correspondences were never systematically repeated in a sequence, not even in the simplified forms (over the same set or by substitution) illustrated in the data-analysis section. Only the building blocks of such operations, that is, single, isolated one-to-many correspondences for multiplication and many-to-one correspondences for division, were shown, but at their most elementary level. Table 7 summarizes the data.

TABLE 6.
Sequences of Inversions on One-Step Series

Age in Months	One-Item Sequences		Two-Item Sequences		Three-or-More-Item Sequences		Total of Inversions	
	N	%	N	%	N	%	N	%
M22	13	68.5	2	31.5	—	—	19	100
M34	6	75	1	25	—	—	8	100
C16	10	40	3	24	2	36	25	100
C36	27	69.2	6	30.8	—	—	39	100
C48	93	44.7	22	21.2	18	34.1	208	100

TABLE 7.
Counivocal Correspondences

Age in Months	1:2 N	1:3 N	2:1 N	3:1 N	Total
M22	—	—	—	—	—
M34	2	—	1	—	3
C16	1	—	—	—	1
C36	1	—	1	—	2
C48	16	3	31	2	52

Counivocal correspondences were virtually absent in macaques and in cebus up to the age of 48 months. At that age cebus presented a striking increase which involved up to 4 objects in both 1-to-n and n-to-1 correspondence.

Only one-to-many counivocal correspondences are produced by 6-month old children. At 8 months, both one-to-many and many-to-one correspondences of minimal size (1:2 and 2:1) are generated, while at 12 months their size extends to four objects (1:3 and 3:1). Repeated counivocal correspondences, never shown by our subjects, begin at 15 months and by 18 month they are constructed on two contemporaneous sets, thus generating real multiplication and division.

Equivalence operations

Biunivocal correspondence

True biunivocal correspondence between two sets of objects was exceptionally produced only once by cebus at 48 months. In fact, this is the remaining case of the 21 cases of two operationally related sets considered in Table 7. The subject constructed a set of two objects by putting one cup on top of another; it then proceeded to construct a second set by putting again one cup on top of another, aligned and oriented in the same way as the first.

Apart from this, the most advanced constructions in this domain were very few instances of sets that appeared partitioned, at least in terms of their component elements, into minimal, symmetrically corresponding subsets. They were generated by both species, but none at the youngest ages. These few constructions are diagrammed in Fig. 9.

Development of biunivocal correspondence in children is comparatively precocious, starting at 10 months.

Simple, symmetric placements of objects within sets, in the form of both horizontal and vertical alignments, were instead produced by both species

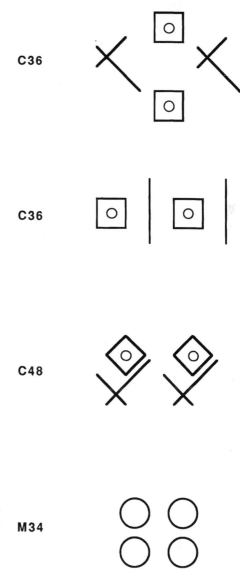

C36

C36

C48

M34

(Three times)

FIG. 9. Schematic representation of the most advanced cases of biunivocal correspondence observed. Each set is composed by four objects but appears partitioned into two corresponding subsets. Symmetry in the orientation of objects is not so exact as depicted.

TABLE 8.
Within-Set Symmetries

Age in Months	Two-Object Sets				Three-Object Sets				
	Horizontal N	Vertical N	Total N	% on 2-Object Sets	Horizontal N	Vertical N	Total N	% on 3-Object Sets
M22	10	—	10	2.2	—	—	—	—
M34	11	6	17	5.2	—	—	—	—
C16	4	4	8	1.9	—	—	—	—
C36	47	18	65	9	5	1	6	3.1
C48	95	46	141	9.3	18	3	21	3.4

at all ages. Table 8 summarizes these data. In the last column of this table sets where symmetries were constructed are reported as a proportion of the total number of two- and three-object sets constructed.

Macaques did not significantly increment their production of symmetries with age. At 22 months only simple two-object horizontal alignments were produced, to which very few vertical ones were added at 34 months. These symmetries were constructed only between two objects.

Cebus monkeys showed a progressive increase with age of frequency and rate of production of symmetries. They were constructed since the earliest age in both the horizontal and vertical dimension. Furthermore from the age of 36 months they extended such constructions to three-object sets: in these cases, correspondence is established between two objects and then reiterated on a third object.

Some developmental progression, thus, appeared to take place within this domain. Any progression, however, stayed within the limits of a single set construction: correspondence was established by both species between single objects and not between sets of objects, apart from the primitive few cases seen above.

Both within-single-set precursors of biunivocal correspondences and true biunivocal correspondence between two sets develop comparatively precociously in children. Both horizontal and vertical symmetric placements of two objects are realized by 6-month-old children. At 8–10 months they become frequent and are extended over three objects. Already at 10 months and regularly at 12 months children produce biunivocal correspondences between two sets or subsets similar to the most advanced types found in our subjects.

Exchange operations

Table 9 reports all occurrences of exchange operations, divided into the three types of replacements, commutativities and substitution, together with their relative proportion of occurrence and the proportion of each type over the total number of compositions generated.

Percentage of exchange operations on total of compositions was stable across ages in macaques, staying at a high fifty percent: this means that about half of all compositions generated were transformed by an exchange operation. Percentage of exchange was almost stable and at about the same level also in cebus, only showing a slight decrease at 36 months.

Commutativities represented by far the most frequent operation in all cases, and in both species their relative proportion with respect to the other two types grows with age. In fact, in 34-month-old macaques they represent almost the only type of exchange operation performed.

On the other hand, the relative distribution of the other two operations shows an opposite trend in the two species. In cebus there is a decrease

TABLE 9.
Exchange Operations: Types

Age in Months	Replacements			Commutativities			Substitutions			Total		
	N	%	% Over Total Comp.s	N	%	% Over Total Comp.s	N	%	% Over Total Comp.s	N	%	% Over Total Comp.s
M22	22	8.3	4	215	80.5	39.2	30	11.2	5.5	267	100	48.7
M34	5	2.3	1.1	212	96.4	49	3	1.3	0.07	220	100	50.2
C16	35	14.8	6.7	197	83.1	38	5	2.1	0.1	237	100	44.8
C36	31	8.6	3	323	89.2	31.6	8	2.2	0.8	362	100	35.4
C48	123	10.2	4.8	1054	87.2	41.6	31	2.6	1.2	1208	100	47.6

of the most elementary operation, replacement, accompanied by a very slight increase (or, at least, stability) of the most complex, substitution. In macaques there is a decrease of both. What is striking and especially important in this case is the disappearance of substitutions.

Substitution, in fact, constructs both the quantitative invariance of a set and the abstract notion of an element as a unit. The fact that they are at all ages few in number and that do not grow with age (on the contrary, they disappear in macaques), a course opposite to that of children, indicates strong limitations toward the construction of any numerical concept of quantity.

Between 6 and 10 months all children produce substitutions in sets of two objects. They increase in frequency and are generated in more and longer sequences.

As shown in the data-analysis section, repeated sequential application of exchange operations can generate various rich structures.

As shown in Table 10, the vast majority of exchange operations were performed in uninterrupted sequences of two or more operations, and not in isolation, by both species at all ages.

The fact that the highest proportion is that of sequences of four or more operations makes clear that invariant transformations of a set are systematically experimented.

However, if we turn to the structural complexity shown by such sequences the picture appears much more primitive than the quantitative data would lead one to suppose.

First of all, notice that the greatest proportion of exchange operations are performed within minimal, two-object sets, as shown in Table 11.

In both species there is a regular decrease of this proportion with age, accompanied by a regular increase in the proportion of larger sets, but even at the oldest ages operations confined to two-object sets are still over 65% of the total.

Children perform all three exchange operations on sets of 3 objects at 12 months. At 15–18 months such operations are applied to sets of three, four and even more objects. Most important, however, is the fact that at the same age children begin to apply these operations to two contemporaneous sets, performing either simultaneous or corresponding operations in both sets. No such cases are found in our subjects.

As discussed in the data-analysis section, important parameters structuring sequences of commutativities and sequences of substitutions are the extent to which all objects of a given set are operated upon and the extent to which successive operations are negation of the preceding ones.

In macaques, at both ages, only about a third of all sequences of commutativities performed on two-object sets invest both objects of the set. Commutativities performed on three or four-object sets never invest all

TABLE 10.
Sequences of Exchange Operations

Age in Months	Isolate Operations		Two-Item Sequences		Three-Item Sequences		Four-or-More-Item Sequences		Total # of Operations	
	N	%	N	%	N	%	N	%	N	%
M22	46	17.2	28	20.9	11	12.4	24	49.5	267	100
M34	34	15.4	12	11	11	15	24	58.6	220	100
C16	48	20.2	28	23.6	12	15.2	13	41	237	100
C36	123	33.9	42	23.2	20	16.5	21	26.4	362	100
C48	234	19.4	99	16.4	62	15.4	90	48.8	1208	100

TABLE 11.
Distribution of Exchange Operations in Relation to Set-Size

Age in Months	On Two-Object Sets		On Three-Object Sets		On Four-or-More-Object Sets		Total	
	N	%	N	%	N	%	N	%
M22	237	88.8	14	5.2	16	6	267	100
M34	169	76.9	34	15.4	17	7.7	220	100
C16	202	85.2	19	8	16	6.8	237	100
C36	264	73	66	18.2	32	8.8	362	100
C48	791	65.5	258	21.3	159	13.2	1208	100

TABLE 12.
Reversibility in Sequences of Commutativities

Age in Months	One Inversion N	Two Inversions N	Three Inversions N	Compensation N	Total N
M22	7	1	—	—	9
M34	8	1	1	—	11
C16	5	1	—	—	7
C36	4	2	1	—	9
C48	15	—	2	3	24

objects, and are limited, in general, to commuting two objects. Cebus show some limited progression along this dimension. In two-object sets both objects are commuted rarely at 16 months, and about 50% of the times at both 36 and 48 months. In three object sets, all three objects are never commuted at both 16 and 36 months, but are commuted at 48 months, although quite rarely. In larger sets of 4 or more objects, commutation of all objects is never performed.

Table 12 reports all cases of sequences of commutativities structured by negation. They are present in both species at all ages, but they are not very numerous.

The most frequent case is that of the simplest single inversion (One inversion), thereby the first commutativity changes the order of one object with respect to the other(s) and the second moves the same object back into its original order relation, thus inverting the effect of the first operation and reestablishing the starting order.

This inversion might be immediately repeated a second time (Two inversions) and even a third time (Three inversions), but as it can be seen, this occurs very rarely (3 times in macaques and 6 in cebus). Furthermore, in the vast majority represented by two-object-set cases, such sequences of inversions are constituted by back-and-forth order-permutation of the same object.

Only cebus at 48 months, in fact, produced the more complex negation by reciprocity where both elements of a two-object set are operated upon; that is, where the original order of the set is changed by displacing one object and then reestablished by displacing the second object in such a way as to compensate the effect of the first commutativity and not simply invert it. Negation by reciprocity was, however, performed only 3 times, as opposed to 21 cases of negation by inversion.

Data on the size of sets where commutativity is applied, on the number of objects that are successively commuted within each set and on inversions performed correspond to the level commutativity operation reaches in children of 12-to-18 months.

Severely limited appears the more complex of the three operations, substitution. Except cebus at 48 months, no other subject ever performed a minimal sequence of two substitutions. Only two sequences, both of the permutational type (xy = C), were shown by cebus at this last age. The first involves three consecutive substitutions:

(1) A B   (2) C B   (3) C A

The second is a sequence of four substitutions, where the last one is also a negation by inversion:

(1) A B   (2) A C   (3) B C   (4) A C

This elementary level of structuring substitution is surpassed even by 10-month-old children.

Negation

As seen above, the basic structure of negation by inversion is present in both species at all ages and applied to addition, subtraction and commutativity.

Macaques appear much more limited than cebus in reversibility on quantitative transformations (bidirectional series). Only cebus at the oldest age shows inversion in substitution, and thus, in this case, negation complements all direct operation present (multiplication and division, in fact, are not even present in their direct form). Again, only cebus at 48 months, shows the more complex negation by reciprocity and only in commutativity. From a structural point of view, negation operations appear strictly limited to the one-step level, i.e. they can only invert the immediately preceding operation and never invert a sequence. There is only one exception to this last statement, which is represented by the unique case, reported above, of a two-step subtraction followed by the inverse two-step addition, also present in 48-month-old cebus.

DISCUSSION

A global consideration of the results obtained by the analysis of logical operations, without considering the respective levels reached by each species, shows two quite different developmental trends.

Macaques do not undergo any substantial development between 22 and 34 months from both a structural and a quantitative point of view. In fact, structures constructed are present since the beginning and remain stable, while their quantitative expression undergoes, in general, a regression rather that an expansion, or, in the most favorable cases, remains stable too. Structural stability characterizes first and second-order series, their inversions constructing bidirectional series, symmetries within sets, type of exchange operations and structure of their sequences, and negation. Quantitative decrease ranges from the number of compositions constructed, to that of overlapping compositions, of bidirectional series constructed, of substitutions performed.

Cebus, on the other hand, show in general a gradual progression in both structural and quantitative development. Both developments appear particularly strong at the 48-month age point. Structural development occurs in simple and bidirectional series, counivocal correspondence, biunivocal correspondence, sequence of exchange operations and in complementing all direct operations by negation.

Quantitative increase occurs strongly, from basic composing to all structures constructed, between 16 to 48 months. Only the intermediate age point, 36 months, shows, in a few cases, some slight quantitative regression or stability with respect to the 16-month age point (number of series and their inversions, but not their structure; proportion of exchange operations performed with respect to compositions, proportion of longest sequences of exchange operations, but, again, not their structure). In no case, however, is there a regression in the structural complexity of the operations involved.

Let's now compare the two species to each other and to children. Level of composing with respect to size of the sets constructed reaches that of 12-month-old (macaques) or 15-to–18-month-old children (cebus), through a quite similar developmental progression. Macaques seem to plateau at a 4-object-set size.

Temporal features of composing evolve, instead, in a quite different way in the nonhuman vs. human primate. Composing two sets in temporal overlap does not go beyond the level of 10-month old children. In macaques, it regresses from this level at the younger age to that of 8-month-old children at the older age, while in cebus it shows the opposite movement. Composing sets in temporal isolation decreases with age as in children: it goes to the 10-month old level in macaques and to the 12-to-15-month old level in cebus. Both species increase their level of composing temporally sequential sets, which instead remains stable in children between 10 and 18 months. These contrasting indexes are the manifestation of an important developmental difference. The proportion of sets constructed in temporal isolation, hence of sets that are not related by some operation to one another, decreases in all species. But while in children from 10 months on it decreases in favor of an increase in the construction of temporally overlapping sets, in nonhuman primates it decreases in favor of an increase in the construction of temporally sequential sets. Now, constructing sets in temporal overlap necessarily results in the production of (at least) two contemporaneous sets, while constructing sets in temporal sequence might have the same result but it also includes all the cases where only one set is constructed and successively transformed into another. This means that the shifting proportion of children directly translates into the opportunity of operationally relating two actual sets, while that of non-human primates might only increase the opportunity of operationally relating successive transformations of a single set. The first can give rise to second-order composing, while the second remains within first-order composing.

In the second place, even when sequential composing does result in the construction of two actual sets, this result is obtained in a way different from that of children.

Children begin to deal with two contemporaneous sets, rather than with one at a time, by constructing them simultaneously and this is the main avenue through which logical operations interrelating two actual sets develop, as shown by their strong increase in overlapping composing between 10 and 18 months. Both macaques and cebus do not follow this course, as shown by the fact that their level of overlapping composing does not increase beyond the initial one of children. To the extent that they interrelate two contemporaneous sets, they do so by constructing them one after the other. This is confirmed by the data reported in Table 3: the vast majority of two contemporaneous sets (137 out of 176, or 78%) are constructed sequentially in spatial proximity but not in temporal overlap and their number strongly increases with age in both macaques and cebus.

This difference has important consequences on the construction of logical operations. Construction of two contemporaneous sets in children originates often from the simultaneous (or partially overlapping) action of identical or similar schemata operating in parallel on objects. Typical examples of such schemata in 10-month old children are reported by Langer (1980, p. 221; lines with identical number denote simultaneous events):

45.5 R[ight] H[and] raises then lowers Green Triangular Column 3 to chair seat
45.5 L[eft] H[and] raises then lowers Red Block 3 to chair seat

10.5 RH drops Green Triangular Column 2
10.5 LH drops Red Triangular Column

86.5 RH brushes Red Block 2 to floor, right side
86.5 LH brushes Red Block 1 to floor, left side

Through these schemata, a simultaneous construction of two sets might easily originate. For example, already at 10 months (Langer, 1980, p. 251):

15.5 RH places Yellow Cross Ring on top of Green Rectangular Ring
15.5 LH places Yellow Rectangular Ring on top of Green Cross Ring

These constructions become common at 12 months. In these cases the symmetry in the construction process tends to produce a symmetry in the result: with some amount of regulation, constructional symmetry results into two sets that spatially correspond element-to-element to each other. And, in fact, first biunivocal correspondences between two sets, or between two subsets are even produced by 10-month-old children and occur at 12 months.

This does not happen in nonhuman primates. Two contemporaneous sets are not constructed simultaneously but separately one after the other. Hence their construction does not tend to be mediated by the occurrence of parallel action-schemata, and does not tend to result into spatially corresponding set.

This differential course of development might explain why the fundamental operation constructing equivalence among sets, biunivocal correspondence, is, even comparatively, strongly underdeveloped in both nonhuman primate species. Only one case of biunivocal correspondence between two minimal two-object sets was constructed as such by Cebus at 48 months. Six more cases, none occurring at the youngest ages in both cebus and macaque, were adjustments of corresponding symmetries within single 4-object sets, dividing it into two subsets.

This contrasts sharply with both nonhuman primate species development in the construction of series. Macaques at 22 and cebus at 36 already construct instances of second order series, even if they are isolated cases and confined to minimal (2- and 3-object) sets. Macaques at 34 extend these constructions to larger sets (3 and 4 objects), though their frequency remains minimal. Cebus at 48 months, on the other hand, constructs a number of these series with even larger sets (4 and 5 objects), reaching a level that surpasses that of 18-month-old children.

On the other hand, construction of second-order series begins in children at 12 months, that is even slightly later than the construction of correspondence between sets (10 months). In fact, Langer (1986) notes that:

[In the second year] the development of ordered nonequivalence by proto-addition and protosubtraction lags behind the progress in structuring equivalence by correspondence, as it does during most of the first year. . . . There is parallel development between first-order series and correspondence. Structuring second-order series, on the other hand, continues to lag developmentally. (p. 81)

This statement can be exactly reversed to describe what happens in nonhuman primates: out of a total of 26 second-order operations relating two sets, 25 are constituted by additions and subtractions and only 1 is a biunivocal correspondence and it occurs only at the oldest age. However considered, in frequency, structural complexity, age of development, correspondence lags way behind series.

It should be noted that the construction of second-order series in children is also mediated, as that of correspondence, by temporally overlapping construction of the two sets (see Langer, 1980, p. 358). And, in fact, this way of construction is responsible for the fact that along their development they tend to be constructed with some form of spatial correspondence between the two sets (like same or similar spatial configuration of the two sets; see Langer 1986, pp. 182–183). In other words, it looks as if ordered nonequivalent sets tend to "grow out" of corresponding equivalent sets, or at least, out of the same construction process. Again, nothing like this happens in nonhuman primates. As already noted in the result section, all constructions of series take place in temporal sequence and no element of spatial regulation or correspondence between the two sets is present.

At the first-order level, a mosaic pattern of similarities and differences characterizes the development of the nonhuman vs. human primates.

Precursors of correspondence, i.e. within-set symmetries, appear comparatively less-developed than in children. In macaques at both ages and cebus at 16 months, they barely reach the level already presented by 6-month-old children. In 36 and 48-month-old cebus they are comparable to those of 8–10-month-old children. From that age on, children, as just seen, begin to develop full correspondences between sets and subsets.

Much the same picture is offered by the comparison of counivocal correspondences. The rare instances presented by macaques (only at 34 months) and cebus at 16 and 36 months parallel those constructed by 8-month-old children (though certainly not in frequence). Cebus at 48 months is instead comparable to children at 12 months. From 15–18 months children begin to construct such correspondences on two sets and generate correspondences of correspondences (second-order correspondences), thus giving origin to real multiplication and division. No repetition, even on the same set is observed in nonhuman primates.

The general backwardness of these structures might be related to the same developmental difference discussed above. The same process of constructing simultaneously by correspondence at an earlier level, where only two objects are manipulated, seems to be responsible for children's early advanced performance in these domains. It is interesting to remember that

Forman (1982) found the origin of symmetrical placement of objects in children in preceding symmetrical action, beginning at 7 months of age.

First-order series, on the other hand, appear comparatively more advanced. In both frequency and structural complexity, macaques at both ages and cebus at 16 and 36 months seem to match children at 12 months. Regularly occurring inversions on such series seems to indicate a minimal level of quantity conservation, i.e. at the level of two and three-element sets. As seen above, there is no development in the two species at the ages considered, but, rather, a quantitative decrease. Cebus at 48 months show a consistent quantitative increase, and some structural development, with occasional three-step series and two-step inversions, which takes them to the level of 15–18-month-old children.

Nonhuman primates exchange operations show a striking split in comparison to childrens'. With the exception of cebus at 48 months, substitutions are barely at the initial level of 6-month-old children. Extended sequences and simple inversions, which are already found in 10-month-old children, only occasionally occur in 48-month-old cebus. Commutativities, on the other hand, in both frequency and structure (size of the sets commuted, number of objects commuted, sequences and inversions) are at the level of 12–15-month-old children, for both macaques and cebus, and at the level of 18-month-old children for cebus at 48 months.

Between 15 and 18 months, however, children begin to apply these operations to two sets and they substitute objects between two sets. No such second-order development appeared in nonhuman primates.

CONCLUSION

Two processes characterize children's development in the logical domain. On the one hand, there is a structural as well as a quantitative increase in the construction of first-order operations, i.e., operations relating single sets, or, more exactly, successive transformations of one set. This leads, for example, to increasing size of compositions, to longer first-order series, to exhaustive commuting and substituting within single sets of increasing dimension, etc. On the other hand, *and at the same time of this first process*, children start constructing (or, one could say, "reconstructing") the same operations at the second-order level, i.e, they begin to operationally interrelate two contemporaneous sets. The beginning of this second process is situated between 10 and 12 months. As in most cases of cognitive developmental structures, construction at a higher level of cognitive functioning runs again through stages analogous to those already passed during development at the lower level (this is why one can talk of "reconstruction";

compare the Piagetian notion of "decalage"): for example, while at 12 months first-order series might relate sets of 4 and 5 objects, second-order series will be constructed on minimal sets of 2 and 3 objects. In the second year of children's life we witness to this parallel double construction in "decalage," and, in fact, first-order construction appears even to slow down, as compared to the second half of the first year. By the end of the second year, however, second-order structure construction is completed, and, "though still not predominant even by age 24 months, infants' constructions of second-order operations begin to balance the advances in their first-order operations" (Langer, 1986, p. 367).

The first important difference characterizing nonhuman primates developing logical structures is the absence of this second parallel developmental process *even in those structures whose first-order development is comparatively advanced and matches that of children in the second year.*

In fact, as we have just seen, the only second-order structures constructed by macaques and cebus up to 36 months add to an insignificant total of 5 minimal series. Cebus at 48 months seem to go further, but their second-order constructions are again confined to one-step series and one instance of biunivocal correspondence.

Hence, one cannot conclude that second-order level is structurally unaccessible to these species of nonhuman primates, but certainly it does not develop at all in its coverage toward "balancing the advances in first-order level". The very evidence of structural stability accompanied by quantitative decrease or stability in macaques shows that these animals have already reached their upper limit in this domain. Cebus developments at 48 months might give the impression that further progress is still possible, but it should be remarked that specific development of second-order structures is practically nonexistent in the whole year that goes from 36 to 48 months: what we see is only a quantitative increase in the number of second-order series produced and the only biunivocal correspondence constructed is a small step away from the only two subset correspondences constructed at 36 months. In other words, there is nothing even distantly approaching children's second-order developmental trend in the course of the second year. The clearest example of this difference is offered by the development of the exchange operation of commutativity.

First-order structures constructed by all our subjects match at least those of children at 12 months and extend to those of 18 months, but in none of the subjects there is any development of second-order structures.

The corollary of this developmental pattern might be the quantitatively more extensive excercise of first-order operation seen at least in cebus at the most advanced age. These results are also in agreement with, and confirm, what was already seen, in the preceding chapter, in the development of classification. Even where classification progressively develops, it remains always within the limit of single sets: in no case two class-consistent sets are constructed.

Absence of parallel development in second-order structuring is, however, not the only difference between human and nonhuman primate construction of logical knowledge. A second fundamental difference lies, as we have seen, in the "overlapping-corresponding" mode of childrens' construction vs. the "sequential-noncorresponding" mode of nonhuman primates. Its developmental result is a specific backwardness, not only in absolute but also in relative terms, i.e., in relation to other developing structures within the same species, in the construction of equivalence operations. Correspondences and substitution lag at the most initial levels shown by children. The developmental consequences of this pattern are substantial.

Quantity-preserving shape and spatial transformations of sets are experimented in commutativities, but they are not complemented by the essential contruction of set-elements as units, i.e. as purely quantitative entities independently of their physical and spatial properties, effected by substitution.

Absence of this last construction coupled to the absence (or very limited) construction of element-to-element biunivocal correspondence makes it very difficult to conceive any possible development toward the construction of the number concept.

15 Comparison with the Human Child

Jonas Langer
Department of Psychology
University of California at Berkeley

This chapter has two comparative goals. One is empirical. The aim is to analyze the major similarities and differences between the cognitive development of nonhuman primates and human infants. The other goal is theoretical. The aim is to continue our discussion and proposals on the evolution of intelligence (Langer, 1988, 1989).

We have previously proposed that cognitive development is a recursive process of multistructural, multilevel, and multilinear change (Langer, 1986, pp. 378–383). Multistructural change means developing different forms of knowledge, including logical (e.g., classificatory) and physical (e.g., causal) cognition. Cognition first becomes multilevel in human infancy towards the end of infants' first year, when first-order operations[1] begin to co-exist with second-order operations.[2] Cognitive change is multilinear in several important respects, including that different structures (e.g., logical and physical) and levels of cognition diverge in their development and that higher levels of cognition hierarchically integrate lower levels of cognition.

There are significant formal similarities and differences between the cognitive development of nonhuman primates and human infants. The fundamental formal similarities between primate species' cognitive development are complex. Nevertheless, the overall picture seems fairly straight-

---

[1]First-order cognitive operations are individual operations that infants map directly onto individual elements, e.g., substituting one object in one set of objects to produce quantitative equivalence over time in the set.

[2]Second-order cognitive operations are operations that infants map onto the product of other operations, e.g., composing two corresponding sets of objects and then substituting objects between the two sets to produce quantitative equivalences upon equivalences.

forward. The findings reported in this volume amply demonstrate that the basic cognitive structures found in human infants are also found in non-human primates. This includes all first-order logical operations, such as classifying single sets only (see page 241). It includes all first-order physical cognition, such as causal dependency (e.g., repeated pushing of one object by another such that the displacements are a direct function of the pushes; see Langer, 1986, pp. 370–375). It also includes other basic categories of physical cognition, particularly of space and objects such as permanence. Thus, all the basic structures of cognitive development are in place in nonhuman primates. Their cognitive development, like that of humans, is multistructural.

This order of shared elementary multistructural cognitive development makes the lack of comparable progress by nonhuman primates toward more advanced intellectual development all the more striking. Neither cebus nor macaques (this volume, Chapters 12, 13, and 14) develop the second-order operations that human infants develop during their second year, let alone the more advanced concrete and formal operations that develop in late childhood and adolescence (e.g., Inhelder & Piaget, 1958, 1964). Rare instances of rudimentary second-order logical but not physical cognition are found in cebus but not macaques (see Chapter 14). Two alternative explanations can plausibly account for why this does not lead to any further cognitive development by cebus. One alternative is that the rare findings may be false positives due to imprecisions in the methods. The other possibility is that rare second-order logical cognition is a necessary but not sufficient condition for further intellectual development. In particular, the crucial missing ingredients for further development may include the lack of concomitantly producing rudiments of the full range of second-order logical and/or second-order physical cognitions.

To begin to understand why the cognitive development of cebus and macaques is so arrested, so minimal compared to humans even during infancy, we will examine central differences between the cognition of these primate species. We will pay close attention to formal differences other than mere divergences in extent of cognitive development. Formal similarities coupled with only formal differences in extent could mislead us into believing that cognitive ontogeny recapitulates cognitive phylogeny (see Langer, 1988, for further discussion; and Parker & Gibson, 1979, for a contrary view).

The fundamental differences revealed by our analyses are, like the similarities, complex in detail, and they go far beyond mere differences in extent. Our analyses are guided by the following proposal. The direction, organization, and sequencing of cognitive development, particularly during its formation in infancy, determines the extent of development within a species. Comparative differences in direction, organization, and sequencing thus determine the relative progress of human over nonhuman cognitive development.

THE DIRECTION OF COGNITIVE DEVELOPMENT

The *elements* (objects) of organisms' (subjects') cognition constrain the level of their intellectual operations. Two factors are crucial. The constancy and power of the elements constrain what kinds of operations are possible. Systematic logical cognition (such as a classificatory system) and physical cognition (such as a system of causal experimenting) is not possible without constant given elements. Further, the more powerful the elements the more they open up possibilities for organisms to develop new and progressively advanced logical and physical knowledge.

Set construction is a major source of cognitive elements for primates (Chapter 14). However, their comparative growth curves diverge (see Figure 1). With increasing age the rate of producing sets: (a) does not change in macaques; (b) increases sharply in cebus; and (c) first increases a bit (during the first year) and then (with some minor fluctuations) remains basically stable (during the second year) in human infancy.

These formal differences in primates' comparative growth curves for set construction are important. They highlight divergence in a fundamental feature of constructing elements for cognition, namely, the rate at which elements are produced. The direction of the growth curves diverge in primates such that their developmental end-points differ. Macaques end up (at age 34 months) producing fewer sets than human infants (at age 24 months). Cebus end up (at age 48 months) producing more sets than either macaques or humans.

Two indices of the constancy and power of the sets constructed are their size (i.e., the number of objects composed) and the temporal relations between the sets. The developmental direction, but not the extent, of set size construction is similar in primate (see Chapter 14, Tables 1a and 1b). Human infants, cebus, and macaques construct increasingly larger sets of objects as they grow older. The main difference is that human infants continue to construct ever larger sets. Eventually, their cognitive elements are much more powerful than those available for the cognition of cebus and macaques. For example, most of the constructions by cebus and macaques at all ages never exceed sets of two objects when presented with six objects. In comparison, by age 24 months most of the sets constructed

Monkey Age (months)

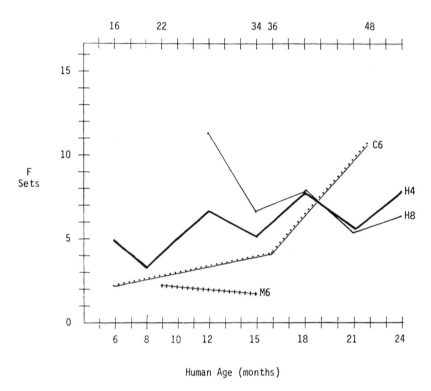

H4 = Humans in four-object conditions
H8 = Humans in eight-object conditions
C6 = Cebus in six-object conditions
M6 = Macaques in six-object conditions

FIG. 1. Mean rate of producing sets per minute.

by human infants comprise three or more objects when presented with eight objects (Langer, 1986, p. 314, Table 15.6).

The developmental direction of the temporal relations between sets is initially similar, but eventually diverges in primates. The beginnings are reflected in Tables 2a and 2b (Chapter 14). With age, the proportion of temporally isolated sets decreases while sequential sets increases in cebus and macaques up to maturity and in humans up to age 18 months. However, the direction of development diverges in human infants after age 18 months. Within 3 months, by age 21 months, the proportion of sequential as well as temporally isolated sets decreases. On the other hand, the proportion of temporally overlapping sets continues to increase (Langer, 1986, p. 241, Table 11.9).

The comparative consequences are twofold. The sets constructed by infants, but not cebus and macaques, continue to become progressively constant given elements because they are increasingly maintained over time in relation to each other. In turn, human infants' constant givens become progressively powerful elements because they increasingly comprise two or more sets of three or more objects in relation to each other. Cebus and macaques are basically limited to cognitions that can be mapped onto single isolated sets of three or fewer objects.

Divergent or partially divergent growth curves usually produce different cognitive end-points. As we have just seen, this applies to sets as elements of cognition. It also applies to the logical and physical *operations* mapped onto these elements. Consider for illustrative purposes the comparative growth curves for constructing causal relations by primates (see Figure 2).

A substantial proportion of the set constructions by young human infants are causal (e.g., propulsion by pushing a block into a cylinder which, as a consequence, rolls away). The proportion increases with age from 6 until 10 months, when almost half of infants' compositions are featured by causality. Afterwards, infants' production of casual relations declines steeply so that by the end of their second year only 1 in 10 sets are causal.

In short, the ratio of producing causal relations by human infants first increases and then decreases with age. The rate of production by cebus hovers between one-fourth and one-third of their set constructions. The data picture for macaques is still incomplete, because data at only two age-points have been taken so far, but the rate of producing causal relations is negligible and the growth curve flat. Thus, the operations of causal cognition as well as their elements end up diverging in these species.

We have taken pains to analyze some independent growth curves on the elements and operations of cognition. The purpose was to show that there is independent data to support the claim that the direction and end-points of the comparative development of both the elements *and* operations of cognition diverge in different primate species. It should be apparent, nevertheless, that once the elements diverge, then the operations that can be mapped onto them must inevitably also diverge. Most important, the constancy and power of the elements constructed by cebus and macaques is limited to first-order elements (e.g., single sets). Cebus and macaques are therefore always constrained, even as adults, to first-order causal mappings onto these elements. It becomes impossible for cebus and macaques to construct second-order causal relations that human infants begin to generate toward the end of their first year. To illustrate, infants begin to construct two temporally and spatially corresponding causal compositions at age 12 months (Langer, 1980, p. 345):

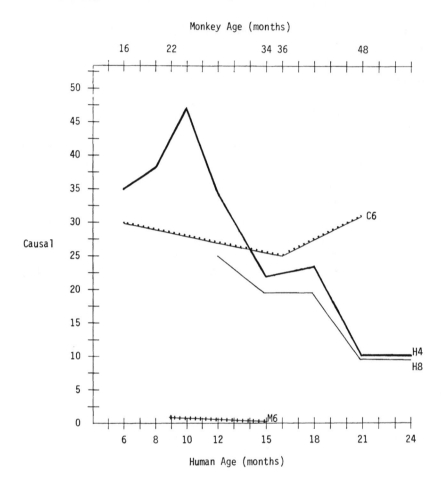

FIG. 2. Percent of sets that are causal.

37.5. RH places Blue Brush 1 next to Red Brush 1 and uses BB1 to push RB1 toward experimenter

37.5. LH places Blue Brush 2 next to Red Brush 2 and uses BB2 to push RB2 toward experimenter

The general conclusion that we draw from such comparative growth curves is that the cognitive development of cebus and macaques is restricted by the elements they can construct. Their development is limited to constructing singular isolated elements, such as small individual sets of objects. The development of human infants is not limited in this way. They progressively construct multiple temporally related elements, such as two corresponding sets of objects. The comparative result is that cebus and macaques are locked into mapping no more than first-order cognitive operations. In contrast, progressive possibilities open up for children to map new and more advanced cognitive operations.[3]

THE ORGANIZATION OF COGNITIVE DEVELOPMENT

Cognitive operations, like the elements onto which they are mapped, are initially constructed by organisms themselves. As the elements progressively approximate constant givens, they open up new and ever-growing possibilities for organisms' operations. These operations map qualitative or intensive (e.g., classifying objects within a set) and quantitative or extensive (e.g., commuting objects within a set) relations (i.e., part-whole transformations). At first human infants' operations are elementary and weak, and should probably be called proto-operations. Progressively they become ever more complex and powerful mappings that increasingly approximate, but never achieve, the status of fully formed logical operations

---

[3]Composing single sets of objects originates, develops, and dominates human infants' constructions during their first, but only their first year. Composing single sets continues to dominate cebus' and macaques' constructions throughout their development. As long as they are limited to constructing single sets, human infants are also limited to mapping elementary operations onto them. So first-order operations mark logical cognition exclusively during most of human infants' first year (but only their first year), and all of cebus and macaques development. With progress in constructing single sets comes progress in the development of first-order operations. First-order operations therefore become increasingly powerful even during human infants' second year, when they no longer exclusively mark logical cognition. Progress in constructing single sets provides progressive elements for increasingly powerful first-order operations by human infants. This, we shall see in section 2, is not always the case for cebus and macaques. Thus, at age 48 months cebus map relatively powerful commuting, but not substituting, onto their more advanced single-set constructions.

Elementary first-order operations, such as substituting and classifying, completely dominate all of cebus' and macaques' cognitive development. While they also dominate human infants' cognitive development during their first year, they no longer do so during human infants' second year. The rudiments of second-order operations originate towards the end of human infants' first year (Langer, 1980) and develop during their second year (Langer, 1986). Fundamental to human infants constructing second-order operations is their forming elements comprising minimal compositions of compositions. One manifestation of this advance is the initiation of composing two sets of objects in temporal overlap during the first year; and by the increase in constructing such contemporaneous collections during human infants' second year. By age 24 months, 20% of infants' productions are temporally overlapping sets in four-object conditions and 33% in eight-object conditions (Langer, 1986, p. 321, Table 15.10).

TABLE 1.
Percent of Human Infants Beginning to Commute, Replace, and Substitute
Two- and Three-Object Sets

Age (mos.)	Commute		Replace		Substitute	
	2-Object	3-Object	2-Object	3-Object	2-Object	3-Object
6	42	0	42	0	33	0
8	92	0	100	0	83	0
10	100	0	100	0	100	0
12	100	42	100	42	100	50

during human infancy. We refer to the former as first-order operations because they comprise direct elementary mappings, and to the latter as second-order operations because they comprise mappings upon mappings.

The organization of first-order operations differs among primate species throughout their development. To illustrate, a set of operations with which organisms can begin to map quantitative or extensive transformations onto sets are exchange operations of substituting, replacing, and commuting (see Chapter 14 for detailed definitions of and scoring criteria for these three operations). Parallel development marks all three operations in human infancy (see Table 1), but not in cebus and macaques. Part of the cognitive organization of cebus development should suffice to make the point (see Chapter 14). At age 48 months, roughly maturity, cebus barely generate the most rudimentary first-order substituting, comparable to those 6-month-old humans begin to construct. At the same time, cebus produce fairly advanced first-order commuting, comparable to those constructed by 12- to 15-month-old humans.

In comparison, all three operations are almost exactly aligned with each other in human infants' behavior. The onset ages when infants first produce commuting, replacing, and substituting are almost perfectly correlated (Table 1). The correlation extends to the power of the three operations. The onset ages are almost perfectly correlated for the origins of mapping all three operations on both two-object and three-object sets.

Parallel development characterizes the onset ages of human infants' first-order (and, in fact, second-order) cognition (Langer, 1980, 1986). All structures are fairly well aligned with each other to form a differentiated and hierarchically integrated organization. For example, when human infants first begin to generate rudimentary first-order operations, then all their first-order operations are rudimentary, including substituting and commuting. Their cognitive structures are coherently organized. In human infants one does not find extreme (onset and developmental) mixtures of rudimentary and advanced first-order cognitions comparable to those found throughout the development of cebus and macaques. In comparison, then, cebus' and macaques' cognitive structures are differentiated but uninte-grated; they are comparatively disorganized.

Differentiated and integrated organization means predominant, not per-fect, correlation between the origins (onset) and development (progress) of cognitive structures in human infants. The contrast is with the disor-ganized (i.e., unintegrated mixture of unaligned) cognitive structures found throughout cebus and macaques development. Further, the distinction is meant to capture the comparative developmental organization of cognitive competence, not performance. For instance, cebus can generate advanced first-order commuting (e.g., inverse commuting of three-object sets) but are incapable of also producing advanced first-order substituting (e.g., inverse substituting in three-object sets). Conversely, human infants who can only generate rudimentary first-order substituting (e.g., direct substi-tuting in two-object sets) are incapable of also producing advanced first-order commuting (e.g., inverse commuting of three-object sets).

In an early investigation, Inhelder (1968) found extreme disparity be-tween the cognitive structures of mentally retarded children. Inhelder hy-pothesized that extreme disparity produces structural "friction" that impedes cognitive development. Our claim is not that cognitive mixture and dé-calage are not found in human infants. Rather, the claim is that disparities between operations (for example, between commuting, substituting, and classifying) are typical but comparatively narrow-gauged. If the principle of reciprocal interweaving (Gesell, 1946) is operative, then the expected developmental consequence is relatively rapid resolution. Gains are made by the lagging operations and, sometimes, by the more advanced operations as well. Accordingly, we have hypothesized that narrow-gauged structural disparities are a major source of optimal structural disequilibrium that, together with structural equilibrium, generates progressive development (Langer, 1969, 1974, 1980, 1986).

A progressive developmental consequence of narrow-gauged structural disparity is that the more advanced operations open up new possibilities for lagging operations. Typical in this regard is how slightly more advanced extensive operations (e.g., substituting) may open up new possibilities for mapping intensive operations (e.g., classifying). By age 24 months most human infants are either on the verge of or already beginning to generate second-order substituting (e.g., exchanging elements between two contem-poraneous sets) and classifying (e.g., composing two contemporaneous and contrasting classes of objects such as a set of 4 circular shapes and a set of 4 square shapes). An important consequence of developing such second-order operations is that human infants try them out in situations for which they were not initially constructed. This opens up myriad possibilities for constructing new and more advanced cognitions. For example, infants be-gin to apply second-order substituting to classifying objects. This includes beginning to correct nonverbal counterconditions posed to them. In one countercondition, we presented human infants with two alignments of 4 shapes each in which one alignment comprised 3 circular shapes and 1 square shape and the other alignment comprised 3 square shapes and 1 circular shape. At age 24 months, some infants correct the classificatory errors by substituting the singular square and circular shapes for each other (Langer, 1986). In so doing, these human infants apply substituting (an extensive operation initially designed to produce quantitative relations of equivalence and nonequivalence) to solving a problem in classifying (qualitative relations of identities and differences). The result is that these infants open up new possibilities for their intensive classifying operations to progress (e.g., to form two contrasting classes).

Our hypothesis is twofold: (a) well-aligned (approaching in-phase) structural organization is a source of progress; while (b) poorly aligned (approaching out-of-phase) structural organization is (at best) not a source of or (at worst) impedes progress. Accordingly, the ever-widening disparity between both cebus' and macaques' operational structures (e.g., between commuting and substituting) as they grow older leads to nonoptimal structural disequilibrium. The probability of inducing mutual structural progress (such as between substituting and classifying found in human infants) decreases as structural disparity increases with age in cebus and macaques.

Structure and development are reciprocal. Extreme structural segregation enhances developmental divergence; and extreme developmental divergence enhances structural segregation. Because it impedes progress, this is a nonoptimal relation between structure and development. Nonoptimal structural development is clearly more characteristic of the cognitive growth of cebus and macaques than of human infants. Moderate structural parity enhances developmental convergence; and moderate developmental convergence enhances structural parity. Because it induces progress, this is an optimal relation between structure and development. Optimal structural development is clearly more characteristic of the cognitive growth of human infants than of cebus and macaques.

THE SEQUENCE OF COGNITIVE DEVELOPMENT

If structure and development are reciprocal; and if the multistructural organization of cognition differs in primate species (as the comparative data have led us to conclude in the previous section); then we may also well expect the temporal sequencing of multistructural development to differ in primate species. Two types of analyses are central to understanding the sequencing of multistructural development. On the one hand, development comprises temporal sequencing within individual structural domains of cognition as they evolve (e.g., within individual physical cognitions such as object permanence and within individual logical cognitions such as set classifying). Within sequence analyses permit us to make elementary analyses of comparative development. On the other hand, development comprises temporal sequencing between all the structural domains of cognition as they evolve, including most importantly between all the structures of physical and logical cognition. Between sequences analyses permit us to make holistic analyses of comparative development.

The comparative data on temporal sequencing *within* individual cognitive structures is beginning to reveal formal differences as well as similarities. Fundamental formal similarities are most evident in sequencing within individual domains comprising physical cognition (e.g., object permanence and causality). Such formal similarities are documented, for example, in Chapters 3 to 10 (this volume). To illustrate, the stage sequence of object permanence first discovered by Piaget (1971a) in human infants is apparently universal (see Doré & Dumas, 1987, and Langer, 1988, for recent reviews). Both the order (no reversals) and closure (no gaps or stage skipping) of the stage sequence are replicated in all the species studied so far (e.g., cats, monkeys, and apes).

Fundamental formal differences in sequencing within individual domains have only been discovered so far in classifying. We will first outline the sequence of developing first-order classifying (i.e., classifying single sets only) in human infants (Langer, 1980, 1986). At age 6 months, infants consistently couple objects from different classes with each other when presented with two contrasting classes of two objects (e.g., 2 identical crosses and 2 identical triangles). For example, 6-month-olds consistently pair crosses with triangles, rather than crosses with crosses or triangles with triangles. At age 8 months, infants no longer consistently couple objects from different classes with each other. Instead, their couplings are random. Thus, for example, 8-month-olds are equally likely to pair crosses with triangles as they are to pair crosses with crosses and triangles with triangles. By age 12 months, infants begin to couple identical objects with each other (e.g., red crosses with red crosses), but only infrequently. Somewhat varying procedures and analyses by Nelson (1973), Ricciuti (1965), Starkey (1981), and Sugarman (1983) yield comparable results on classifying by identities at age 12 months. By age 15 months, infants begin to couple consistently similar (e.g., red with blue crosses) as well as identical (e.g., red with red crosses) objects with each other.

Compare the sequencing of developing classifying by cebus and macaques (reported in Chapter 13). In cebus the sequence is from (a) mostly random classifying and partially classifying by identities and similarities (at age 16 months), to (b) mostly classifying by differences and partially classifying by identities and similarities (at age 36 months), to (c) classifying by identities and similarities (at age 48 months). Only a two-step sequence is reported for macaques: (a) mostly classifying by identities and similarities and partially classifying by differences (at age 22 months), to (b) classifying by identities and similarities (at age 34 months).

The comparative development of classifying is unusual because the sequences diverge (are formally different), yet the cognitive end-points converge (are formally similar). The three species take divergent developmental paths towards eventually classifying by identities and similarities. This comparative finding is clearly inconsistent with a recapitulation account.

So far, then, the comparative developmental findings indicate (a) formal similarities within sequencing of physical cognitions, coupled with (b) formal differences within sequencing of logical cognitions. We do not yet have sufficient data to determine with confidence that this is a general comparative pattern. But if it turns out be general, then it will have major import for understanding the evolution of intelligence. Among other ramifications, it would indicate that: (a) sequences that are physically necessary (such as object permanence) are universal; while (b) sequences that are logically arbitrary (such as classifying) are not universal even if their developmental end-points are.

The comparative data on temporal sequencing *between* logical and physical cognition point toward fundamental formal differences. Logical and physical cognition develop simultaneously and in parallel in human infants (Langer, 1980, 1986). Physical cognition (such as of object permanence and of causal relations) develops at the same time and in synchrony with logical cognition of intensive (e.g., classificatory) and extensive (e.g., equivalence by substituting) relations.

In contrast, central physical cognitions develop before logical cognition in cebus and macaques. The development of these physical cognitions is well underway or completed by the onset of logical cognition. Most striking, cebus complete their development of object permanence (which extends through stage 5, that is, the acquisition of presentational object permanence) as young as age 7 months (see Chapters 7 and 8). Yet cebus only begin to develop logical cognition during their second year (see Chapters 12 to 14).

The data picture on developing the physical cognition of causality is a bit more complicated. Developing simple causal means-ends relations (such as using a support as an instrument to obtain a goal object) is completed no later than age 9 months in cebus and 15 months in macaques (see Chapter 9). On the other hand, developing more complex causal means-ends relations (such as using a stick as an instrument to obtain a goal object) may not be completed until ages 18 to 20 months in cebus; and it may never develop in macaques (see Chapter 10). Thus, the development of elementary causal cognition is well underway (in macaques) or completed (in cebus) by the onset of logical cognition. The development of more advanced causal cognition is well underway (in cebus) or nonexistent (in macaques) by the onset of logical cognition.

These comparative differences indicate that developmental sequencing between cognitive domains diverges in primate evolution. Logical cognition begins to develop after physical cognition in cebus and macaques ontogeny. The development of their physical cognition is well underway by the onset age at which they begin to develop logical cognition. Logical cognition originates at the same age (i.e., early infancy) as physical cognition in human ontogeny. Further, the development of their physical cognition is in progress at the same ages at which their logical cognition is in progress. Neither type of cognition begins or ends before the other during childhood. Hence, both are open to similar environmental influences and to each other's influence.

In the early development of cebus and macaques, physical and logical cognition constitute consecutive developmental trajectories that are relatively independent of each other. Asynchronous developmental trajectories do not readily permit direct interaction or information flow between cognitive structures because they are out of phase with each other. At most, then, the interaction between cognitive domains in the early development of cebus and macaques may be indirect. The main potential lines of influence are from relatively developed physical cognition to undeveloped logical cognition.

From the start of human ontogeny, physical and logical cognition constitute contemporaneous developmental trajectories that become progressively interdependent. Synchronic developmental trajectories facilitate direct interaction or information flow between cognitive domains. Mutual and reciprocal influence between logical and physical cognition is readily achievable because they develop simultaneously and in parallel. Thus, we have found (Langer, 1985; 1986, pp. 375–378) that even in infancy, logical cognition introduces elements of necessity and certainty into physical cognition. At the same time, physical cognition introduces elements of contingency and uncertainty into logical cognition.

CONCLUDING HYPOTHESIS

Development from unilinear growth (of physical cognition) to multilinear growth (of physical and logical cognition) in the ontogeny of cebus and macaques is "folded over" in primate evolution to form descendant multilinear development (of logical and physical cognition) from the start in human ontogeny. The onset age for beginning to develop physical cognition is roughly the same in all primates (as far as we know). But the onset age for beginning to develop logical cognition is accelerated in humans.

Phylogenetic displacement in the ontogenetic onset or timing of one feature relative to the other developmental features within the same organism causes a disruption in the repetition of phylogeny in ontogeny. This type of evolutionary change was called heterochrony by Haeckel as reinterpreted by deBeer (see Gould, 1977, for exhaustive discussions of the history of heterochrony). Because it involves a dislocation of the phylogenetic order of succession, heterochrony produces a change in the rate or timing of ancestral processes. The timing may be accelerated or retarded. Thus, heterochrony is an evolutionary mechanism by which ancestral correlations between growth, differentiation, and hierarchic integration (i.e., correlations in orthogenesis)[4] are disrupted and new descendent orthogenetic correlations are established. In a word, the pattern of orthogenesis (and all that entails for ontogenesis) is altered.

The comparative data on the organization of and sequencing between cognitive domains are consistent with the hypothesis that heterochrony is a mechanism of the evolution of primate intelligence. On this, our central evolutionary hypothesis (that requires further comparative testing, especially with chimpanzees), heterochrony is the mechanism whereby consecutively developing ancestral cognitive structures were transformed in phylogenesis into simultaneously developing descendant cognitive structures in human ontogenesis. Heterochrony produced the reorganization of nonaligned ancestral cognitive structures found in cebus and macaques into the comparatively well-aligned descendant structural development of cognition found in human infancy (analyzed in section 2). This (orthogenetic) reorganization opened up the possibility of information flow between (logical and physical) cognitive structures in human infancy. These same domains are segregated from each other in time, and, therefore, in information flow in the early development of nonhuman primates such as cebus and macaques.

Foundational formal differences in organization, sequence, and direction of cognitive development are at the heart of the minimal intellectual ontogeny of nonhuman as compared with human primates. As already noted, we have proposed that human cognitive development is the synthetic product of both equilibrated and disequilibrated structures. In comparison, the relatively disorganized and asynchronic structures of cebus and macaques indicate a predominance of disequilibrium over equilibrium. The synethetic opportunities for opening up new cognitive developments are therefore minimized or closed off in the ontogeny of cebus and macaques. In comparison, the equilibrium and disequilibrium conditions of the organization, sequencing, and direction of humans' recursive (i.e., multi-structural, multilevel, and multilinear) cognitive ontogeny maximizes continuing and open-ended intellectual progress.

---

[4]We follow Werner's (1948) proposal that the orthogenetic process of development is directed toward increasing differentiation, centralization, and hierarchic integration.

V CONCLUSION

16 On the Phylogeny of Human Cognition

Francesco Antinucci
Istituto di Psicologia, C.N.R
Rome, Italy

Ever since Haeckel formulated his famous biogenetic law that ontogeny is the short and rapid recapitulation of phylogeny, it has been extraordinarily hard to resist, in any field of comparative study, the recurrent temptation to frame results of comparisons between "higher" and "lower" organisms, especially when they involve ontogenetic development, in terms of recapitulation central mechanism: that of "terminal addition" (see Gould, 1977). New features of descendant species are added in evolution to the end of the ontogeny of ancestral species: the repetition of this process at each successive step generates the ontogenetic recapitulation seen in descendant species.

Consequently, ontogenetic development in more advanced species reruns and then goes beyond that of less advanced species. The comparative study of cognition, in as long as it has been concerned with evolutionary questions, has been no exception to this tendency. One has only to go through Doré and Dumas' (1987) review of piagetian studies of animal cognition, for example, to see how pervasive this position is. Piaget's proliferation of developmental subdivisions into periods, stages, substages, etc. (often used for sake of exposition, since no strict temporal correlate or internal correspondence is attached to them) has easily lent itself to a direct reification in phylogeny. Thus, cognition in nonprimates reaches between stage 2 and stage 4 of sensorimotor intelligence, depending on the species, monkeys go up to stage 5 or 6, while great apes also develop representational intelligence, maybe confined to the early preoperational period. The cognitive developmental sequence seen in the human child recapitulates its phylogenetic history and, at its end, adds new and more advanced stages to it. We see here the hallmark of recapitulation: what are adult structures in lower species correspond to transitional ontogenetic structures in higher ones.

While some differences are found, not in one instance are they discussed as potentially implying that an altogether different structure through a divergent developmental course is being constructed. To quote just one example: "The role of interactions with physical objects is more limited in great apes than in humans, and circular reactions (primary, secondary and tertiary) are displayed in fewer modalities and contexts" (Doré & Dumas, 1987, p. 223). These differences tend to be dealt with as "special adaptations", in a way strongly reminiscent of how Haeckel himself dealt with "exceptions" to his biogenetic law: "The true and complete repetition of phyletic development by biontic [ontogenetic] development is falsified and changed by secondary adaptations" (Haeckel, 1866, quoted in Gould, 1977, p. 81). We shall not repeat here the fundamental critiques that have been moved against the recapitulation thesis, based on both theoretical and empirical arguments. Two of them are, however, directly relevant to our discussion.

One is the empirical finding that the second mechanism necessary to recapitulation, beside terminal addition, "condensation" or acceleration of development, is often absent, and this holds particularly true in the case of our data.

> The higher conditions have been produced by a crowding back of the earlier characters and an acceleration of growth, so that a given succession in order of advance has extended over a *longer range of growth* than its predecessor *in the same allotted time.* (Cope, 1870, quoted in Gould, 1977, p. 87)

Quite the opposite pattern characterizes our data: development, far from becoming more accelerated in the higher "conditions" (as Cope puts it), shows the inverse trend, as seen in chap. 6. If anything, retardation of development characterizes at least the earlier phases of human development with respect to the other primate species in both absolute and relative terms.

The second argument is von Baer's (1828) old but brilliant one that similarities among ontogenies in related species arise not because of recapitulation but because of general common processes of embryonic development (his "laws of development"). Since development proceeds universally by increasing differentiation, more specific structures will always follow more generic ones. This produces two main observable consequences: on the one side, commonality will be automatically greater the earlier the developmental phase compared; on the other side, a sequence of developmental stages within a given domain will be necessarily the same if, and in so far as, each stage of the sequence is a prerequisite for the differentiation of the following one. One should not wonder, therefore, at similarities determined by these general processes, and construe them as empirical findings.

Thus, as we have seen in chapters 3–6, stage 1 of sensorimotor intelligence is identical in all the species considered, while from stage 2 to stage 4 differences among the same species progressively increase. The developmental sequence of object-concept, which is a typical case where each stage depends on the achievements of the preceding one (for example, objects cannot be conceived as occupying only one place in space unless they have first been constructed as detached from the subject's action), follows the same course in all species. On the other hand, the order of development of schemata intercoordinations in stage 2, where no such dependence exists, is open to the possibility of following different paths in each species (and, in fact, so it does).

Let's, then, "recapitulate" our findings and see how, at all levels where difference is manifested, human cognition is in no way the product of monkeys' or apes' cognitive development that has gone one (or more) step(s) further (or, conversely, that monkey's or apes cognition cannot be characterized in terms of earlier ontogenetic stages of human cognition plus some "special adaptations").

A first divergence already occurs early in development. Nonhuman primates do not develop the physical domain of cognition parallel to the logical domain of cognition as human infants do. While, up to the fourth stage of sensorimotor development, the structures built in the logical domain are essentially the same in all species, those built in the physical domain do not match. The largest difference is that of the human vs. the nonhuman primates, but, within this domain, nonhuman primates species also differ with each other. We saw, furthermore, that this difference does not match the phyletic difference among the species: apes do not come closer to humans than monkeys do. If we attempt to trace it to its origin through the operation of the functional constants constructing knowlege (according to the program outlined in the introduction), we find early differences in the nature of the organizing action on the environment. Basically, two modes of interacting with external objects are found: a direct (or "primary") one through body movements and an indirect (or "secondary") one through intermediaries. Their different availability, and, subsequently, the different proportion in which they are exploited, in the different species, appears to be related to an intricate temporal interaction among stage durations, development of prehension-coordinations and development of independent locomotion, as outlined in chap. 6.

Contrary to the prediction of recapitulation, a general and pervasive pattern of retardation characterizes, from this point of view, the development of the human infant with respect to the other primate species; one that, in fact, has been found also in several other dimensions (see Gould, 1979, for a review).

But the most important factor seems to have been a specific "heterochrony", in de Beer's (1930) sense: the temporal displacement in the ontogenetic sequence of the time of onset of locomotion capacity with respect to that of the other capacities. The comparison shows that this has been relatively pushed forward to an exceptional degree in the human infant. Again contrary to recapitulation prediction, it is the New World monkey cebus, and not the ape, that comes second on this dimension, though at quite a distance. Gorilla and macaque then follow in the order.

As it may be expected from the structural nature of cognitive constructions, these early differences are amplified by further development; they do, however, maintain their initial proportions. At all successive stages, human children's construction of the physical domain by far outstrips that of any other species, but, again, however tested (stick problem, object manipulations), cebus comes second, followed by gorilla and macaque.

A second divergence, and, insofar as one can tell, one which is completely unrelated to the first one, occurs at a later stage of ontogeny: it concerns the development of representation. Here the pattern of relationship among species is different. Both macaque and cebus do not appear to develop a capacity for mental representation (in the piagetian sense, as defined in the introduction), as the human infant does toward the end of his sensorimotor development. The gorilla, on the other hand, gives evidence to be able to construct at least rudimentary forms of mental representation. As judging from our results (see chapters 3 and 8), probably enough to construct elementary symbolic-reference relations. This seems to be confirmed by studies of the language-related capacities of this species, though, as we shall see below, satisfying the representational prerequisite is still a far cry from the possibility of constructing a full-fledged language capacity.

Already at this point, it should however be clear that even if nonhuman primates "went on" to build a full representational capacity, as children do, their resulting cognition, though certainly more powerful than their actual one, would be structurally different from that of children.

And yet, another set of independent divergences is still to come. They relate, this time, not to the initial, but to the more advanced construction of the domain of logical knowledge. Two main differences are found.

All basic types of unit that structure logical cognition in the human infant are also found in the nonhuman primate species. However, after constructing first-order logical structures, but, most importantly, while they are still developing them in scope and extension, children begin a parallel construction of second-order structures (for example, operations whose elements are sets themselves, rather than elements of a set, or classifications where two sets of similar objects are simultaneously constructed). This development is not found in the nonhuman primate species. Both the data on logical operation (see chap. 14) and those on classification (see chap. 13) show that second-order structures (and, hence, arguably, all higher-order structures) are not constructed along developing first-order structures. In fact, except from sporadic and rudimentary isolated instances, that do not undergo any structural development, they are never constructed.

Notice again that this too is not a case of arrested (or prolonged) development within the same dimension: children *do not* construct second-order structures *after* they have completed first-order ones. Rather their construction, at a given point, "branches off", and proceeds in a parallel, but asynchronous, fashion on the two lines: second-order structures, in fact, retrace, at the higher level, the initial development of first-order one. Nonhuman primates' development is instead unilinear: their going on along the same line would not generate human cognition.

This is not the only difference found in the development of logical cognition. A second, important one, is in the *mode* of constructing, even at the level of first-order structures. As seen in chap. 14, children develop very early an "overlapping-corresponding" mode of construction by means of contemporaneous and parallel action schemata which tends to produce spatial symmetries and correspondences in the resulting constructions. Nonhuman primates constructions, on the other hand, are almost exclusively produced by temporally sequential constructing, even in those cases where two actual sets are generated. Consequently, in these constructions no spatial symmetry or correspondence is automatically introduced by the action schemata generating them. At its origin, this appears to be again a fundamental difference in the nature of the organizing action. Contrary to the one seen above, unfortunately, we are not able, in this case, to indicate its "proximate cause". It may be, however, that even the absence of second-order constructing is dependent on it: at least in so far as parallel constructing in children initiates and facilitates the production of two sets which tend to be, by its very process, intrinsically regulated with each other.

The differences just seen in the development of logical cognition give the final blow to any picture of cognitive evolution as a unilinear process. Advancing the development of the logical constructions showed by non-human primates would not produce the logical structures of human cognition.

The consequences are, obviously, widespread: consider language as an example. Of the many prerequisites for developing human language capacity, three are relevant in this context. As a system of symbols, language requires a representational capacity. Furthermore, these symbols have to be structurally organized in at least two ways. On the one side, they have to be organized into classes, some of which must be "open", i.e., capable of infinitely expanding their constituency in a regulated fashion. This presupposes a capacity for rule-based, recursive, classification. On the other side, such classes must be combined into constituents in a hierachically growing fashion (the linear order of symbols in a sentence is the product of an all-encompassing hierachical syntactic structure). This process requires the capacity of constructing sets of sets, sets of sets of sets, and so on recursively, again in an open fashion, and of structuring, at each successive level, relations among them. Though it cannot be excluded in principle, the empirical evidence of our studies on both classification and logical operations militates against the eventual development of the first of these two capacities: class and set construction, far from openness, does not even reach exhaustiveness in constituency. Presence of the last capacity, instead, can be excluded in principle: it cannot, obviously, be developed within the limit of first-order logical constructions.

To state the point more directly, then, even if our nonhuman primate species would develop a capacity for representation, fundamental components of a language capacity, as seen in humans, would still be unaccessible to them. Unfortunately, we could not study systematically, as we did in cebus and macaque, the development of logical structure in our only ape species, the gorilla. Yet, insofar as we can informally infer from the capacities displayed in other tasks (for example, in the stick problem), its logical abilities do not seem to parallel those of children anymore than those of the other species do.

If this inference is correct, and can be generalized to all ape species, it might provide a straightforward explanation of the specific limitations found in apes' development of language-like abilities. Since they appear to have at least an elementary representational capacity, apes should be able to deal with a system of symbols (even if they do not spontaneously produce one, which is a separate question). Their structuring such a system would, however, be severely constrained. On the one hand, they could not form real rule-based lexical classes. This would surface as a continued slow and laborious expansion of their lexicon, as compared to the rapid and accelerating expansion rate that characteristically steps in at a certain point of child language development. Second, and most important, given the limitation on higher-order constructions, their sentences could not be structured by more than a single level of hierachical constituency, which, for example, would exclude the generation, within the same sentence unit, of productive (but not fixed) constituent-modification.

It seems to us that, where accurate and reliable accounts of these indexes of the language abilities of apes have been reported (for example, Terrace et al., 1979), they tend to agree with these predictions.

It is worth noting that a distribution of capacity converse to that seen in apes could also be hypothesized. That is, absence of representational capacity but presence of higher-order logical structuring. Though there would be no language, cognitive-dependent activities would be much more human-like than in apes. One might wonder whether the production of symmetrical implements (of the so-called Acheulian variety), appearing at a given point of the human evolutionary sequence (possibly, with *Homo erectus*), might not just signal the establishment, or, at least, the beginning point of such development.

In conclusion, the evolutionary path leading to the structuring of human cognitive capacities, as we see it today, far from being the progressive furthering of a linear process, seems to have taken several independent "turns" at various steps of its long course. The first appears to have been the one leading to the elaboration of the physical domain. This seems to have occurred, within an ongoing process of progressive retardation of development, through a specific and particularly strong heterochronic displacement. Then, at least two other turning points can be discerned: the one leading to representational capacity and the one leading to the advanced structuring of logical cognition.

Unfortunately, we are, at present, unable to determine both their sequence and their triggering mechanism. Knowing, however, what to look for will offer an invaluable guide to a future research that should span continuously, as it did in the past, from animal cognition to paleoanthropology.

References

Antinucci, F. (1981). Studi cognitivi sui primati non umani. *Giornale Italiano di Psicologia*, 8, 211–240.

Antinucci, F. (1982). Modelli "atomistici" e modelli "strutturali": Il ruolo dei paradigmi nella controversia innatismo/ambientalismo. *Storia e Critica della Psicologia*, 3, 1, 79–105.

Antinucci, F. & Visalberghi, E. (1986). Tool use in *Cebus apella*: A case study. *International Journal of Primatology*, 7, 349–362.

Baba, M., Darga, L., & Goodman, M. (1980). Biochemical evidence on the phylogeny of Anthropoidea: In R. L. Ciochon & B. Chiarelli (Eds.) *Evolutionary biology of New World monkeys and continental drift* (pp. 423–443). New York and London: Plenum Press.

Baer, K. E. von (1828). *Entwicklungsgeschichte der Thiere: Beobachtung und Reflexion.* Konigsberg: Borntrager.

Beck, B. B. (1976). Tool use by captive pigtailed macaques. *Primates*, 17, 301–310.

Beck, B. B. (1980). *Animal tool behavior.* New York: Garland STPM Press.

Bever, T. G. (1974). The interaction of perception and linguistic structures: In T. Sebeok (Ed.) *Current trends in linguistics, vol. IX*. The Hague: Mouton.

Boesch, C., & Boesch, H. (1983). Optimization of nut-cracking with natural hammers by wild chimpanzees. *Behaviour*, 83, 265–286.

Chevalier-Skolnikoff, S. (1977). A Piagetian model for describing and comparing socialization in monkey, ape and human infant: In S. Chevalier-Skolnikoff & F. Poirier (Eds.) *Primate bio-social development: Biological, socialand ecological determinants* (pp. 159–187). New York: Garland Press.

Chomsky, N. A. (1959). A Review of B. F. Skinner's Verbal Behavior. *Language* 35, 26–58.

Chomsky, N. A. (1975). *Reflections on language.* New York: Pantheon Books.

Ciochon, R. L., & Chiarelli, A. B. (1980). Paleobiogeographic perspectives on the origin of the Platyrrhini: In R. L. Ciochon & B. Chiarelli (Eds.) *Evolutionary biology of New World monkeys and continental drift* (pp. 459–494). New York and London: Plenum Press.

Cooper, L. & Harlow, H. (1961). Note on a cebus monkey's use of a stick as a weapon. *Psychol. Rep.*, 8, 418.

Corrigan, R. (1981). The effects of task and practice on search for invisibly displaced objects. *Developmental Review*, 1, 1–17.

de Beer, G. R. (1930). *Embriology and evolution*. Oxford: Clarendon Press.

Dewsbury, D. A. (1984). *Comparative psychology in the twentieth century*. Stroudsburg: Hutchinson Ross.

Dresher, K. & Trendelenburg, W. (1927). Weiterer beitrag zur intelligenz-prufung an affen (einschliesslich anthropoiden). *Z. Verglag Physiol.*, 5, 613–642.

Doré, F. Y. & Dumas, C. (1987) Psychology of animal cognition: Piagetian studies. *Psychological Bulletin*, 102, 219–233.

Eldredge, N. & Gould, S. J. (1972). Punctuated equilibria: An alternative to phyletic gradualism: In T.J.M. Schopf (Ed.) *Models in paleobiology* (pp. 82–115). San Francisco: Freeman, Cooper & Co.

Epstein, R. (1987). Comparative psychology as the praxist views it. *Journal of Comparative Psychology*, 101, 249–253.

Fischer, K. W. & Jennings, S. (1981). The emergence of representation in search: Understanding the hider as an independent agent. *Developmental Review*, 1, 18–30.

Ford, S. M. (1986). Systematics of the New World Monkeys: In D. R. Swindler & J. Erwin (Eds.) *Comparative primate biology, Vol. 1: Systematics, evolution and anatomy* (pp. 73–135). New York: Alan Liss Inc.

Forman, G. E. (1982). A search for the origins of equivalence concepts through a microanalysis of block play: In G. E. Forman (Ed.) *Action and Thought: From sensorimotor-schemes to symbolic operations* (pp. 97–135). New York: Academic Press.

Fossey, D. (1972). Vocalization of the mountain gorilla (*Gorilla gorilla beringei*). *Animal Behavior*, 20, 36–53.

Fossey, D. (1979). Development of the mountain gorilla (*Gorilla gorilla beringei*). The first thirty-six months: In D. A. Hamburg and E. R. McCown (Eds.) *The Great Apes* (pp.138–184). Menlo Park, California: The Benjamin/Cummings Publishing Co.

Freese, C. (1977). Food habits of white-faced capuchins (*Cebus capucinus*) in Santa Rosa National Park, Costa Rica. Brenesia, 10/11, 43–56.

Gardner, B. T., & Gardner, R. A. (1971). Two-way communication with an infant chimpanzee: In A. M. Schrier & F. Stollnitz (Eds.) *Behavior of nonhuman primates, vol. 4.* New York: Academic Press.

Garstang, W. (1922). The theory of recapitulation: A critical restatement of the biogenetic law. *Journal of the Linnean Society of London, Zoology*, 35, 81–101.

Gartlan, J. S. (1974). The African forests and problems of conservation: In S. Kondo and M. Kawai (Eds.) *Proceedings of the 5th International Congress of Primatology* (pp 509–524). Tokio: Japan Science Press.

Gesell, A. L. (1946). The ontogenesis of infant behavior: In L. Carmichael (Ed.) *Manual of child psychology*. New York: Wiley.

Gesell, A. L. & Amatruda, C. S. (1951). *Developmental diagnosis: Normal and abnormal child development*. New York: Hoeber.

Goodall, A. G. (1974). *Studies on the ecology of the mountain gorilla (Gorilla gorilla beringei) of the Mt. Kahuzi-Biega region (Zaire) and comparisons with the mountain gorilla of the Virunga Vulcanoes*. Unpublished Ph. D. dissertation, Liverpool University.

Gottlieb, G. (1987). The developmental basis of evolutionary change. *Journal of Comparative Psychology*, 101, 262–271.

Gould, S. J. (1977). *Ontogeny and phylogeny*. Cambridge: Harvard University Press.

Groves, C. (1986). Systematics of the great apes: In D. R. Swindler & J. Erwin (Eds.) *Comparative primate biology, Vol. 1: Systematics, evolution and anatomy* (pp. 187–217). New York: Alan Liss Inc.

Haeckel, E. (1866). *Generelle morphologie der Organismen*. Berlin: Georg Reimer.

Harlow, H. (1951). Primate learning: In C. Stone (Ed.) *Comparative psychology* (pp. 183–238). New York: Prentice-Hall.

Harlow, H., & Settlage, P. H. (1934). Capacity of monkey to solve patterned string tests. *Journal of Comparative Psychology*, 18, 423–435.

Hobhouse, L. T. (1926). *Mind in evolution*. London: Macmillan.

Hobhouse, L. T. (1968). *Social evolution and political theory*. Port Washington: Kennikat Press. (Original work published 1911).

Hoffstetter, R. (1980). Origin and deployment of New World monkeys emphasizing the southern continent route: In R. L. Ciochon & B. Chiarelli (Eds.) *Evolutionary Biology of New World Monkeys and Continental Drift* (pp. 103–122). New York and London: Plenum Press.

Inhelder, B. (1968). *The diagnosis of reasoning in the mentally retarded*. New York: John Day. (Original work published 1943).

Inhelder, B., & Piaget, J. (1958). *The growth of logical thinking from childhood to adolescence*. New York: Basic Books. (Original work published 1955).

Inhelder, B., & Piaget, J. (1969). *The early growth of logic in the child*. New York: Norton & Co. (Original work published 1959).

Izawa, K. (1979). Foods and feeding behavior of wild black-capped capuchin *(Cebus apella)*. *Primates*, 20, 57–76.

Izawa, K. (1980). Social behavior of the wild black-capped capuchin *(Cebus apella)*. *Primates*, 21, 429–453.

Izawa, K. & Mizuno, A. (1977). Palm-fruit cracking behavior of wild black-capped capuchin *(Cebus apella)*. *Primates*, 18, 773–792.

Jolly, A. (1972). *The evolution of primate behavior*. New York: The Macmillan Co.

Kavanagh, M. (1983). *A complete guide to monkeys, apes and other primates*. London: Jonathan Cape.

Kluver, H. (1933). *Behavior mechanisms in monkeys*. Chicago: Univ. of Chicago Press.

Knobloch, H. & Pasaminick, B. (1959). The development of adaptive behavior in an infant gorilla. *Journal of Comparative and Physiological Psychology*, 52, 699–704.

Kohler, W. (1976). *The mentality of apes* (2nd ed.). New York: Liveright. (Original work published 1917).

Langer, J. (1969). Disequilibrium as a source of development: In P. H. Mussen, J. Langer, & M. Covington (Eds.) *Trends and issues in developmental psychology*. New York: Holt, Rinehart & Winston.

Langer, J. (1974). Interactional aspects of mental structures. *Cognition*, 3, 9–28.

Langer, J. (1980). *The origins of logic: 6 to 12 months*. New York: Academic Press.

Langer, J. (1985). Necessity and possibility during infancy. *Archives de Psychologie*, 53, 61–75.

Langer, J. (1986). *The origins of logic: One to two years*. New York: Academic Press.

Langer, J. (1988). A note on the comparative psychology of mental development: In S. Strauss (Ed.) *Ontogeny, phylogeny, and historical development*. Norwood, NJ: Ablex.

Langer, J. (1989). Early cognitive development: Basic functions: In C.A. Hauert (Ed.) *Developmental psychology: Cognitive, perceptuo-motor, and neuropsychological perspectives*. Amsterdam: North Holland.

Luckett, W. P. (1980). Monophyletic or diphyletic origins of Anthropoidea and Hystricognathi: Evidence of the fetal membranes: In R. L. Ciochon & B. Chiarelli (Eds.) *Evolutionary biology of New World monkeys and continental drift* (pp. 347–368). New York and London: Plenum Press.

Mathieu, M. & Bergeron, G. (1981). Piagetian assessment of cognitive development in chimpanzee *(Pan troglodytes)*: In B. Chiarelli & R. S. Corruccini (Eds.) *Primate behavior and sociobiology* (pp. 142–147). Berlin: Springer.

Mathieu, M., Bouchard, M. A., Granger, L. & Herscovitch, J. (1976). Piagetian object permanence in *Cebus capucinus*, *Lagothrica flavicauda* and *Pan troglodytes*. *Animal Behaviour*, 24, 585–588.

Mathieu, M., Daudelin, N., Dagenais, Y., & Decarie, T. (1980). Piagetian causality in two house-reared chimpanzees (*Pan troglodytes*). *Canadian Journal of Psychology*, 34, 179–185.

Miller, G. A., Galanter, E., & Pribram, K. H. (1960). *Plans and the structure of behavior*. New York: Holt, Rinehart and Winston.

Napier, J. R. (1970). *The roots of mankind*. Washington: Smithsonian Institution Press.

Napier, J. R. & Napier, P. H. (1967). *A handbook of living primates*. New York: Academic Press.

Nellman, H. & Trendelenburg, W. (1926). Ein beitrag zur intelligence-prufung niederer affen. *Z. Vergl. Physiol.*, 4, 142–200.

Nelson, K. (1973). Some evidence for the cognitive primacy of categorization and its functional basis. *Merrill-Palmer Quarterly*, 19, 21–39.

Nute, P. E., & Mills, K. A. (1986). Patterns of macromolecular evolution in the primates: In D. R. Swindler & J. Erwin (Eds.) *Comparative primate biology, Vol. 1: Systematics, evolution and anatomy* (pp. 277–297). New York: Alan Liss Inc.

O'Donnell, J. M. (1985). *The origins of behaviorism. American Psychology, 1870–1920*. New York: New York University Press.

Parker, C. E. (1968). The use of tools by apes. *Zoonooz*. 41, 10–13.

Parker, C. E. (1969). Responsiveness, manipulation and implementation behavior in chimpanzees, gorillas and orang-utans: In C. Carpenter (Ed.) *Proceedings of the second international congress of primatology* (pp. 160–166). Basel: Karger.

Parker, C. E. (1974a). Behavioral diversity in ten species of nonhuman primates. *Journal of Comparative and Physiological Psychology*, 87, 5, 930–937.

Parker, C. E. (1974b). The antecedents of man the manipulator. *Journal of Human Evolution*, 3, 493–500.

Parker, S. T. (1977). Piaget sensorimotor series in an infant macaque: A model for comparing unstereotyped behavior and intelligence in human and nonhuman primates: In S. Chevalier-Skolnikoff & F. E. Poirier (Eds.) *Primate biosocial development: Biological, social and ecological determinants* (pp. 43–112). New York: Garland Publishing.

Parker, S. T. & Gibson, K. R. (1977). Object manipulation, tool use and sensorimotor intelligence as feeding adaptations in cebus monkeys and great apes. *Journal of Human Evolution*, 6, 623–641.

Parker, S. T. & Gibson, K. R. (1979). A developmental model for the evolution of language and intelligence in early hominids. *Behavioral and Brain Sciences*, 2, 367–408.

Parker, S. T. & Potí, P. (in press). The interaction between sensorimotor intelligence and fixed action patterns in the development of intelligent tool use in tufted capuchin monkeys: In S. T. Parker & K. R. Gibson (Eds.) *Language and intelligence in monkeys and apes: Comparative developmental perspectives*. Cambridge: Cambridge University Press.

Passingham, R. E. (1975). Changes in the size and organization of the brain in man and his ancestors. *Brain Behav. Evol.*, 11, 73–90.

Piaget, J. (1951). *Play, dreams and imitation*. New York: Norton. (Original work published 1945).

Piaget, J. (1971a). *The construction of reality in the child*. New York: Ballantine Books. (Original work published 1937).

Piaget, J. (1971b). *Genetic epistemology*. New York: Norton.

Piaget, J. (1972). *Psychology and epistemology*. New York: The Viking Press. (Original work published 1970).

Piaget, J. (1974). *The origins of intelligence in children*. New York: International University Press. (Original work published 1936).

Piaget, J. (1976). *The grasp of consciousness*. Cambridge, Mass.: Harvard University Press. (Original work published 1974).

Piaget, J. (1977). *The Development of Thought: Equilibration of Cognitive Structures*. New York: The Viking Press. (Original work published 1975).

Piaget, J. & Inhelder, B. (1962). *Le developpement des quantites physiques chez l'enfant* (2nd edition). Neuchatel: Delachaux & Niestle.

Piaget, J. & Inhelder, B. (1967). *The child's conception of space.* New York: Norton. (Original work published 1947).

Piaget, J. & Szeminska, A. (1941). *La genese du nombre chez l'enfant.* Neuchatel: Delachaux & Niestle.

Pilbeam, D. (1984). The descent of Hominoids and Hominids. *Scientific American,* 250, 3, 60–69.

Premack, D. (1976). *Intelligence in ape and man.* Hillsdale: Lawrence Erlbaum.

Redshaw, M. (1975). Cognitive, manipulative and social skills in gorillas: Part II, the second year. *Ann. Rep. Jersey Wildl. Pres. Trust.* 12, 56–60.

Redshaw, M. (1978). Cognitive development in human and gorilla infants. *Journal of Human Evolution,* 7, 133–141.

Ricciuti, H. (1965). Object grouping and selective ordering behavior in infants 12 to 24 months old. *Merrill Palmer Quarterly,* 11, 129–148.

Roitblat, H. L. (1987). *Introduction to comparative cognition.* New York: W. H. Freeman & Co.

Romanes, G. J. (1883). *Animal intelligence.* New York: Appleton.

Rumbaugh, D. M. (Ed.) (1977). *Language learning by a chimpanzee: The Lana project.* New York: Academic Press.

Rumbaugh, D. M. & McCormick, C. (1967). The learning skills of primates: A comparative study of apes and monkeys: In D. Stark, R. Schneider and H. J. Kuhn (Eds.) *Progress in primatology* (pp. 289–306). Stuttgart: Gustav Fischer.

Sarich, V. M., & Cronin, J. E. (1980). South American Mammals: Molecular systematics, evolutionary clocks, and continental drift: In R. L. Ciochon & B. Chiarelli (Eds.) *Evolutionary biology of New World monkeys and continental drift* (pp. 399–421). New York and London: Plenum Press.

Schultz, A. H. (1968). The recent hominoid primates: In S. L. Washburn and P. C. Jay (Eds.) *Perspective on human evolution* (pp. 122–195). New York: Holt, Rinehart and Winston.

Schwartz, J. H. (1986). Primate systematics and a classification of the Order: In D. R. Swindler & J. Erwin (Eds.) *Comparative primate biology, Vol. 1: Systematics, evolution and anatomy* (pp. 1–41). New York: Alan Liss Inc.

Shepherd, W. (1910). Some mental processes of the rhesus monkeys. *Psychol. Monogr.,* 7, 1–61.

Shurcliff, A., Brown, D., & Stollnitz, F. (1971). Specificity of training required for solution of a stick problem by rhesus monkey (*Macaca mulatta*). *Learn. Motiv.,* 2, 255–270.

Sinclair, H., Stambak, M., Lezine, I., Rayna, S., & Verba, M. (1982). *Les bebes et les choses.* Paris: Presses Univ. de France.

Skinner, B. F. (1956). A case history in scientific method. *American Psychologist,* 11, 221–233.

Snyder, D. R., Birchette, L. M. & Achenbach, T. M. (1978). A comparison of developmentally progressive intellectual skills between *Hylobates lar, Cebus apella* and *Macaca mulatta*: In D. J. Chivers & J. Herbert (Eds.) *Recent advances in primatology, Vol. 1: Behavior* (pp. 945–948). New York: Academic Press.

Stanley, S. M. (1979). *Macroevolution. Pattern and Process.* San Francisco: W. H. Freeman & Co.

Starkey, D., (1981). The origins of concept formation: Object sorting and object preference in early infancy. *Child Development,* 52, 489–497.

Struhsaker, T., & Hunkeler, P. (1971). Evidence of tool using by chimpanzees in the Ivory Coast. *Folia Primatologica,* 15, 121–219.

Struhsaker, T. & Leland, L. (1977). Palm-nut smashing by *Cebus apella* in Colombia. *Biotropica*, 9: 124–126.

Sugarman, S. (1983) *Children's early thought: Developments in classifications*. Cambridge, Mass.: Cambridge Univ. Press.

Sugiyama, Y. (1981). Observations on the population dynamics and behavior of wild chimpanzees at Bossou, Guinea. *Primates*, 20, 513–524.

Terrace, H. S., Petitto, L. A., Sanders, R. J. & Bever, T. G.(1979). Can an ape create a sentence? *Science*, 206, 891–900.

Thomas, R. K. & Walden, E. L. (1985). The assessment of cognitive development in human and nonhuman primates: In E. S. Watts (Ed.) *Nonhuman primate models for human growth and development* (pp. 187–215). New York: Alan Liss.

Thorington, R. W. (1967). Feeding and activity of Cebus and Saimiri in a Colombia forest: In Stark, Schneider & Kuhn (Eds.) *Progress in Primatology* (pp. 180–184). Stuttgart: Fisher.

Tolman, C. W. (1987). Comparative psychology: Is there any other kind? *Journal of Comparative Psychology*, 101, 287–291.

Torigoe, T. (1985). Comparison of object manipulation among 74 species of nonhuman primates. *Primates*, 26, 2, 182–194.

Uzgiris, J. C. & Hunt, J. McV. (1975). *Assessment in infancy: Ordinal scales for psychological development*. Urbana, IL: University of Illinois Press.

Vaughter, R. M., Smotherman, W. & Ordy, J. M. (1972). Development of object permanence in the infant squirrel monkey. *Developmental Psychology*, 7, 34–38.

Visalberghi, E. (1986). The acquisition of tool use behavior in two Capuchin monkeys groups (*Cebus apella*).*Primate Report*, 14, 226–227.

Vygotsky, L. S. (1962). *Thought and Language*. Cambridge, Mass.: MIT Press. (Original work published 1934).

Wake, D. B., & Larson, A. (1987). Multidimensional analysis of an evolving lineage. *Science*, 238, 42–48.

Warden, C., Koch, A. & Fjeld, H. (1940). Instrumentation in cebus and rhesus monkeys. *J. Genet. Psychol.*, 40, 297–310.

Wasserman, E. A. (1984). Animal intelligence: Understanding the minds of animals through their behavioral "ambassadors": In H. L. Roitblat, T. G. Bever and H. S.Terrace (Eds.) *Animal cognition* (pp. 45–60). Hillsdale: Lawrence Erlbaum.

Watson, J. (1908). Imitation in monkeys. *Psychol. Bull.*, 5, 169–178.

Werner, H. (1948). *The comparative psychology of mental development*. New York: International Universities Press.

Westergaard, G. C., & Fragaszy, D. M. (1987). The manufacture and use of tools by Capuchin monkeys (*Cebus apella*). *Journal of Comparative Psychology*, 101, 2, 159–168.

Wise, K. L., Wise, L. A. & Zimmerman, R. R. (1974) Piagetian object permanence in the infant rhesus monkey. *Developmental Psychology*, 10, 429–437.

Wood, S., Moriarty, K. M., Gardner, B. T. & Gardner, R. A. (1980). Object permanence in child and chimpanzee. *Animal Learning and Behavior*, 8, 3–9.

Yerkes, R. (1916). The mental life of monkeys and apes: A study of ideational behavior. *Behav. Monogr.*, 3, 1–145.

Yerkes, R. (1927a). The mind of a gorilla. *Genet. Psychol. Monogr.* 2, 1–193.

Yerkes, R. (1927b). The mind of a gorilla: Part II. Mental development. *Genet. Psychol. Monogr.* 2, 377–551.

Author Index

Subject Index

Printed in the United States
by Baker & Taylor Publisher Services